The Mythology of Science and Evolution of its Metaphor

Roger Watson Smeeth

2nd Edition

Copyright © 2014 by Roger Watson Smeeth

All rights reserved. No part of this publication may be reproduced, stored in a retrieval system, or transmitted, in any form or by any means, electronic, mechanical, photocopying, recording, or otherwise, without the written prior permission of the author.

ISBN 978-1502383884

rwsmeeth@islandnet.com

Typeset in *Palatino* at SpicaBookDesign

Printed in Canada
by Printorium Bookworks, Victoria B.C.

*Dedicated to
Michael Littrel,
Mythologist,*

whose understanding of the works of Joseph Campbell
forms the foundation of the story of science herein.

Table of Contents

INTRODUCTION... vii

WHY THIS EXPLORATION?............................. 1

THE PLATONIC VIRUS 7

THE ROLE OF METAPHOR IN SCIENCE 9

BEGINNINGS IN GREECE 11

FOLLOWING GREECE 19

INTRODUCTION TO RELATIVITY ETC................. 36

THE MANIFOLD APPROACH........................... 38

ALBERT EINSTEIN..................................... 41

QUANTUM THEORY................................... 55

CHAOS THEORY....................................... 70

CHAOS – A DIFFERENT APPROACH TO MEASURE..... 79

SOME EARLY CHAOLOGISTS 86

CHAOS APPLIED 106

CHANGING THE METAPHOR......................... 109

RECOGNIZING CHAOS 116

CONCLUSION 120

READING LIST....................................... 125

Introduction

Demanding a causal universe, Science studies nature's laws, not the law of nature. Nature's laws are particular metaphors of the overriding Law of Nature which for science is unknowable and in philosophy is described as God. That fundamental distinction hopefully shapes an underlying theme of this exploration.

As you proceed through it, I suggest you bear in mind what I see as three fundamental ideas that may encourage you to exercise this cautionary principle.

Firstly, there is no essential difference between what is a scientist's theory, or myth, and his metaphor. As that metaphor changes, continually attempting to align his myth with his ritual, so does his world view.

The second idea concerns what is termed fisher information (I), which acknowledges that we cannot know what Nature knows about a system – everything. Fisher information is that part that can be extracted from a system, and depending on the degree that the system is seen as closed, could be a tiny percentage. This information is a function of the error inherent in the measurement system – a case of the uncertainty from which flow all the laws of physics except chaos, and which seriously impedes science's search for truth. So, how does it work? Observation of a system acts as a catalyst, the result of which manifest differential equations. The most useful of these are called LaGrangians, interpreted as natures laws that obey "Principles of Least Action and External Distance" – for example, the shortest or most efficient distance between two points, and as a parallel, between a myth and its ritual, linked by its metaphor and describing the degree of authenticity.

The third idea is contained in the Gödel Incompleteness Theory, which essentially states that one cannot use the tenets of a system to prove or disprove itself. One must go outside, using the tenets of another system. This is particularly true in the case of formal mathematics, proving itself mathematically. The validity of this theory has been demonstrated in the century–long attempts to unify Relativity and Quantum, using their respective tenets. Superstring Theory, being "outside", does currently offer the possibility of proposing the "Pot of Gold" – a Grand Unified Field Theory.

Why this Exploration?

I have a friend, a scientist, with whom over, the past ten years, I have had weekly lunch on conversations – our cares for the world, our cases and our passions. One might suggest that my profession, Architecture has a bit of applied science in its engineering but I am definitely not a scientist. I enjoyed my high school chemistry, the linear process of determining the amount of output from a particular chemical reaction. Apart from that experience, I have had little interest in science and how it parses our world. There was never any fascination.

As our wonderful conversations progressed, I began to see how different we were in the way we processed our world realities. I was beginning to experience the fundamental distinction mentioned at the star. Although my friend showed great interest in the thoughts and ideas that stemmed from my intuitions, for him I was talking an almost foreign language. As a scientist, he wasn't able to subject a parallel source within himself to analysis or provide measurable or replicatable evidence. I believe it was that experience that determined me, from my position of virtual ignorance, to explore science and the development, over the years, of its methodology. What results is a story of how well metaphor works as illusion and little about the law of Nature.

The story that has been concocted is my metaphor of an equally regular series of conversations with another friend – Mike, who piqued my interest with a list of those scientists he considered to be the main characters in the development of the Scientific Method, and beyond. My curiosity was focused and the following list, with a few added philosophers, structured my exploration:

Anaximenes	585-524 BCE
Pythagoras	582-507 BCE
Paramenides	515-460 BCE
Socrates	469-399 BCE
Democritus	460-370 BCE
Plato	427-347 BCE
Aristotle	384-322 BCE
St. Thomas Aquinas	1225-1274 CE
William of Ockham	1285-1349 CE
Nicholas of Cuza	1401-1464 CE
Nicholas Copernicus	1473-1543 CE
Francis Bacon	1561-1626 CE
Johannes Kepler	1571-1630 CE
Galileo Galilei	1564-1642 CE
René Descartes	1596-1650 CE
Isaac Newton	1642-1727 CE
James Maxwell	1831-1879 CE
Albert Einstein	1879-1955 CE

Nails Bohr and	1885-1962 CE
Ewin Schŕödinger	1887-1961 CE
Werner Heisenberg	1901-1946 CE
Edward Lorenz	1917-2008 CE
Mitchel Feigenbaum	1944
Benoit Mandelbort	1924-2010 CE
Otto Rossler	1940

Over the last 2500 years, the development of Western Science has been involved in a debate between two systems of study – the Dialectic and the Scientific Method – each of which proposed the limited value of the other. Relying on specific rules and producers, they are both closed systems. In order to make any meaningful comparison of these two systems and adhere to Gödel's Incompleteness Theory, we cannot use the tenets of either system, we would have to find a platform both outside and embracing both. However, as we shall see, the two are so diametrically opposite there is little chance for agreement. Without that bridging platform (which could be a new understanding of Nature itself) the debate, though low key, will go on. A search for "truth" is common to both methods witness the search for the GUT.

The Dialectical Process – (Rational Analysis) The addition of vowels to the Greek alphabet in the 8th century (BCE) caused a major change in the ability to communicate. With fewer symbols being required represent sounds, the ability to tell the previously orally based stories in print was greatly facilitated. Plato, a student of Socrates, in his "Dialogues"and particularly "The Republic", fixed the stories in print. He codified the rules of logic and the authority of writing, creating and absolutism in the written word as a way of eliminating morality as a legitimate vehicle for knowledge. Learned from Socrates, he used the dialectical appropriate vehicle.

The established steps in the process were:

- A thesis is proposed and supported
- To counter it, an anti-thesis is proposed
- Using dialogue and the rules of logic, a synthesis is developed – which in turn becomes a new thesis, and the process continues

Eventually an hypothesis, for the moment satisfactory, is settled on – "Truth" is found until the next challenge. One can see this as a never-ending process. And long as the rules of logic are adhered to there is a degree of openness in this closed system – any question can be considered – even the number of angels on the head of a pin. The discipline comprises logical rational analysis of intellectual and mystical insight, a process of recollection that led the thinker from the shifting shadows to changeless realities, to essences, the eternal invariant characters that constitute the forms of the governing concept of good. How is that for philosophy! Plato contended that the hypothesis was only a point of departure into a world of principle – beyond all hypothesis. And yet, others of the time saw the process as being far too subjective, almost speculative imagination. What was the value of counting those angels? Without objective observation and measurement, how could they expect to discover any credible truth? The seeds of the scientific method were planted to soon root science in the world of matter.

The Demonstrative Process – (Scientific Method) Prior to the above development, a core belief of the science was expressed in the proposal by Pythagoras that the fundamental element of the universe was "numbers"and could be seen, using observation and measurement, in ratio and proportion. The resulting truth could be accessed using Rational Mathematical technique to unearth the order. This belief became part of the foundation of the developing opposition to the Dialectic taking form in the steps of the Scientific Method:

- Observation and measurement of physical events using both senses and technology
- Rational analysis, including dialectical, of resulting data to form an hypothesis.
- Testing the hypothesis by replicating the results through experiment. (Unlike for Plato, the hypothesis was the end point)

Each step in the process was based on a belief system. Data were selected on that basis- that bias. Its method always involved a set of rules and procedures for proper use of the senses in gaining information that could be believed, in the context of that particular belief. To insure that information was valuable, the goal was to remove all human subjective influence from the findings. Given the nature of humans, this is a big challenge. The act of observation must remain innocent. However, "looking"may not be mirrored in "seeing". It is possible for seeing to be modified when it is through the observer's lens of metaphor, thereby losing that innocence.

The fundamental belief of science is that we live in a casual universe. Not only is it imperative those causes be found, it is believed possible, within the bounds of science, to discover them. I feel that very often the search is driven by a conviction, some might say, that borders on speculation, as to where the answer should be found. It is a passion – "We know its there and we'll find it."

The Platonic Virus

When first exposed to this idea, I began to understand more clearly my experience with the lunchtime friend. The basic position represented in the Scientific Method stems from a mythology, a belief system, developed in the 4th century B.C.E., by Plato, the Greek philosopher. Believing that truth existed, and could be known, he held then that everything learned through the senses and believed to be real, wasn't. The senses could not be trusted. Their information was merely "shadows on the cave walls". Only known to the senses it was essentially illusion. He further held that only route to truth involved a journey from the apparency of illusion, from the treeness to the trees, a journey employing intellectual intuition and mystical insight, thus leading to a result that would supplant feelings – a conclusion that one could "think feelings". Like a virus this position has pervaded science to the present, even though it was recognized that intuition and insight were not thoughts, not of the intellect. A corollary of this position is seen in the statement, "No man is more a slave than he who thinks himself free". Of course, freedom is a feeling.

Plato further contended that there are no new thoughts, just recollections. Using the dialectic, it is possible to recollect that which one is not aware of – the only legitimate route to truth but which appears to be a closed loop. Ignoring the unseen, the subjective, his journey to truth was stuck. It could not incorporate what Vedic scriptures referred to as "things that the tongue has not soiled with a word."

Science still remains a prisoner of its own making – a system that owes much to the development of the dialectic. Originally, having no faith in observation based on sensory information, Plato used the dialectic to rationally establish hypotheses and thereby remove the influence of subjectivity. As the Scientific Method evolved into its established process of observation, rational analysis, and experimental verification, it split off from the purely Platonic approach. The dialectic, not being a scientific process, was only used as part of the analysis.

The Role of Metaphor in Science

The list of scientists will be considered in term of the following model:

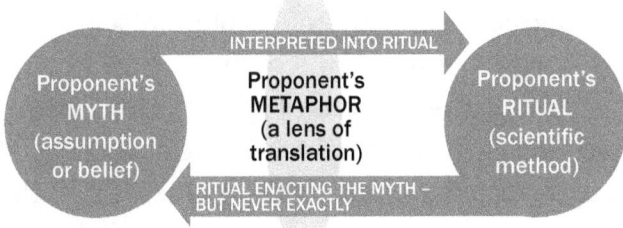

The metaphor is the lens through which the Myth is translated into Ritual, which in turn is enacted into the Myth.

The function of a lens is to focus and narrow the field of vision. The proponent's individual metaphorical translation of the Myth is unique to his or her particular psycho/physiology. Through a parallel process, the lens influences the perceived results of experimental verification in trying to enact the myth – "looking"(observation) and "seeing (assessment) are never identical. Noting the differences makes an evolution of the Metaphor possible, the only way of adding to the Myth.

In using the term "evolution", I do not refer to the 18th century "evolutionist", pre-Darwinian church supported belief in the unrolling of a divine plan towards an ordained goal. I do refer to the post – Darwinian meaning – an ongoing cause and effect development towards some non-predetermined capacity – simply change – a case of entelechy in the realization of potential and not teleology. Darwin was reluctant to use the term "evolution" in his 19th century, "Origin of the Species". Apart from his disagreement with the evolutionists, he was also a product of new burgeoning business culture. His scientific observations of nature proposed three phases in its development – Reproduction, Mutation and Natural Selection. Corollaries for all three can be found in the business world, and when the public translated his idea into "survival of the fittest," it was a perfect allegory for "success of the bottom line". Which came first can be argued, but the connection between science and the context of its culture, good or bad, cannot, and hence the once respected theories of Cultural, Economic and Social Darwinism.

Beginnings in Greece

In this matter of contextual influence, I wondered what wider conditions existed at the time of birth of western science in 6th century BCE Greece. Two centuries earlier Greece had begun its revival from its dark age. Overseas trade was renewed with the resulting great exposure to oriental mysticism, mythology, animism, polytheism, the austere religion of Yahveh in the near east and finally Egypt. This latter culture also had developed a significant relationship between astronomy and mathematics and an understanding of the place of their world in the cosmography. Also, concurrent with this new beginning, written language, which up to then had a very limited capacity, made a huge leap forward with the Greek addition of vowels. Not only did this produce an alphabet that allowed sounds to be represented with fewer symbols, it most importantly allowed a reductionist approach to writing, a necessary capacity for the coming science.

By the six century BCE, Persia was the dominant power in the region and had overwhelmed the Greek Lydian Kingdom on Asia Minor. By this time the greek city states were established and although often at war with each other, they joined forces to repulse two Persian invasion attempts early in the next century. In spite of the vast territory eventually controlled by the Persian Empire, Greece was never occupied or lost its independence, until Rome took control in 147 BCE.

Maybe it was this independence, translated into an independence of thought, that created a transformation of the esoteric, supernatural mythology they saw in the greater region then, into the more rational approach to knowledge that characterized their Golden Age. Mathematics had to be part of that position. Certainly the development of their advanced alphabet helped them fix their new rationalism in print, creating an absolutism of "Truth"in the written word. Of course, part of that "Truth"included the limiting roles of their blossoming rationalism.

The selected seven principal orchestrator s of this blossoming lived a total span of roughly 250 years as diagrammed. After this burst of energy it took more than 18 centuries for the search to be reinvigorated. In the 6th century BCE, the citizens of the city states began to think in new ways. They could ask philosophical questions without recourse to the ancient myths, centred on gods in human form, clothes, and language similar to their own.

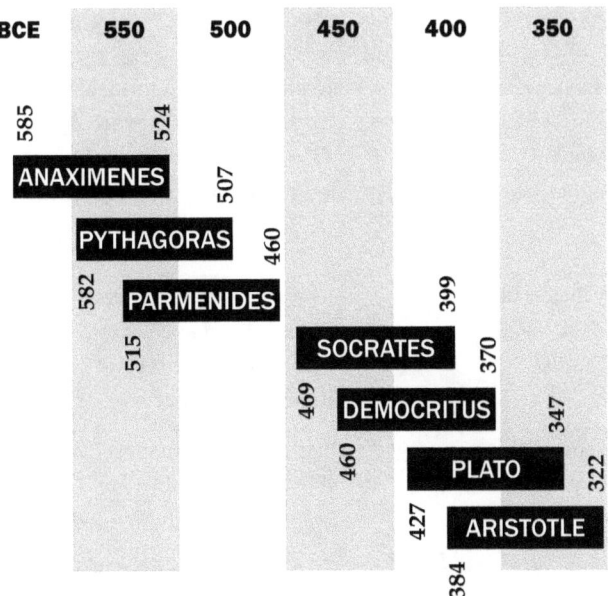

The fundamental development of Greek philosophy was to move from a mythological basis of thought to one based on experience and reason, thus setting the stage for the scientific method. These early Greeks sought to find natural, rather than supernatural explanations of natural processes. Hence the early philosophers, Theles, Anaximander and Anaximenes are referred to as the Natural Philosophers. The latter, and his contemporary Heraclitus, helped to lay the grounds for the fundamental debate between science and philosophy. Anaximenes said that our sensory perceptions were completely unreliable while Heraclitus said, watching the flow of nature, that they were quite reliable.

Beginnings in Greece

PROPONENT	MYTH (his truth)
Anaximenes 585-524 BCE "Monist " Student of Anaximander	Everything is quantifiable. All life, matter and qualities are stages of different aspects of a continuum of which the principle constituent is air.
Pythagoras 582-507 BCE "Dualist"	Everything (the universe) is Numbers – eternal and everlasting. The principle of natural law governs – it is Ratio and proportion.
Parmenidies 515-460 BCE "Monist"	Truth exists as matter. Everything that exists has always existed. The appearance of change is illusion.
Socrates 469 – 399 BCE "Dualist"	Truth exists and man is the measure. Never declare positively what one doesn't know.
Democtitus 460 – 370 BCE "Monist" Inspired by Leucippus	Combinations of elementary and indestructible elements of matter form all there is. (from Parmenidies)
Plato 427 – 347 BCE "Dualist" Disciple of Socrates	Since numbers exist and are absolute, there there must be absolute truth – even though he postulated the opposite. Believed the earth went around the sun.
Aristotle 384 – 322 BCE "Dualist" Student of Plato	Everything – is all matter is animate, an essential spirit called "quality"that can not been seen. Natures process is pulled to final cause – perfection. Time is infinite and the universe was eternal

METAPHOR (How its done)	RITUAL (the doing)
Air could be transformed into all forms and substances thru condensation into cloud water and earth and rarefaction into fire. Diversity comes from different density of air.	Thru reductionism to approximate the qualitative thru the quantitative. He explained the eclipse of the sun.
Truth can only be known thru measurement of ratio and proportion all understandable thru reason.	The truth is accessed with Rational mathematical technique.
All creation exist as a giant sphere – the ideal shape – non changing.	His metaphor didn't work – it could not be verified. Therefore no ritual.
One gets to resolution thru rational analysis.	His dialectic methodology and his syllogism.
All words composed of tiny indivisible spheres (atoms) of matter – not etherial numbers. Man and mud are the same, just different arrangement.	A dialectical investigation of nature in terms of matter – creating a visual representation of numbers.
Truth can only be known thru intellectual recollection of unformed ideas using intellectual and mystical insight – there no new thoughts.	The Socratic dialectic methodology.
Truth is achieved through logic. He created Nichomachian Ethics – foundational element in the scientific method. Postulated the "Great Chain of Being" – from rocks to man.	Observation of nature and use of his syllogism and deductive logic.

Though this debate was temporarily resolved, the question of reliability of sensory information will emerge time and time again.

The following table of our Greek philosopher scientists begins to demonstrate the relationship between Myth, Metaphor and Ritual forward to page 10 – the goal being "Truth".

Given the wide range of myths and ideas proposed by these philosopher/scientists, one overriding belief was common to them all – "Truth existed and it could be discovered". Right to the present in fact, whether or not appropriate, this fundamental belief has driven sciences pursuits. Whatever its particular focus, whatever its chosen process, dialectical, demonstrative, or chaos, the conviction has persisted. Exploring the unknown, the goal has been to verify what was speculatively believed to be the way things should be – that particular "Truth".

When considering the individual Greek proponents in the table, an important distinction should be noticed – the classification as to "Dualism" and "Monism".

Dualist – recognizes two independent principals, matter and mind. There is what is seen and what is not. The latter is fundamental.

Monist – denies the existence and duality of matter and mind. There is nothing other than matter, i.e. there is no separation between man and mud.

This polarity, subjective vs. objective, quality vs. quantity, certainly limited both and made any useful dialogue between the two positions virtually impossible. However, it did set the stage for the two major methodologies of science previously described. Also note that without the technology for refined observation and measurement beyond that of simple aspects, the Greek "scientist", both dualist and monist, relied on logic and dialectic to make their points. What Science has come to require as proper experiential conditions had to wait. For philosophy however, it was an amazing time.

This concentration of thinkers of both persuasions seemed to share that one underlying goal – the discovery of "truth"and the right conditions for its discovery. Although the monists limited their investigations to the material world, both positions relied on rational analysis and it was the development of this methodology that supported the evolutions of science. Basic to this development was a growing respect and honour for what should characterize the goal – "truth". Parmenides translated "truth"into "unexpectedness". He had no preconception of what the goal might be. Socrates wondered if anyone had the right to say positively what one didn't know.

In his "Republic", Plato contended – "Let us agree that [philosophical minds] are lovers of all true being. They should also possess truthfulness – they will never intentionally receive into their false-hood, which is their detestation " – and also – "he whose mind is fixed on true being has his eye always directed towards things fixed and immutable. Can a man help imitating that with which he holds reverential converse? "

Aristotle held that what one said must be consistent with what one thinks – a very common sense philosophy. He pressed the difference between knowledge and opinion, the latter only reflecting temporary truths, if that. Also he held that observation, the foundation of the scientific method, only increased probability but never results in certainty. His production of the first book on logic became the basis for that method. His contention that everything had an essential spirit he called "Quality"set him apart from most of his peers.

During this period, the evolution of philosophy and its concern with understanding the human condition was profound. Knowledge gained thru this discipline could be used to direct lives – individual and whole societies. In this role it is much more useful than science. However, it must be acknowledged that in its search for the ideal, true being, Philosophy has seldom been an obvious driver of culture. Its role has been qualitative, not the quantitative that seems to rule. idealism, faced with what is termed "practical reality", generally suffers rationalization and compromise. Science calls it "renormalization".

Following Greece

One has to wonder why, after this period of concentrated and diligent explorations of Greek science, 1800 years passed before the field was again approached with such passion. What happened to the knowledge and its records during this hiatus?

Rome conquered Greece in the 2^{nd}C, BCE and though her academic traditions were respected – wealthy families had Greek slaves as teachers and many texts were translated into Latin – the new centre of power wasn't interested in theories of pure science. Usefulness was its focus. The knowledge was translated into its application – engineering. Rather than the human condition, Roman philosophy was primarily concerned with the pragmatic means of strengthening the Republic, and later the Empire.

Though of little interest to the Roman establishment, St. Augustine, 354 – 430 CE breathed new life into the philosophical landscape. He was an absolute monist, contending that all meanings originate with God. Valid reasoning was created by God. Man could only observe God's processes of rhetoric and dialectic to discover the meaning of his scriptural intent. Augustine's position was a precursor of the growing absolutism of the church and sciences insistence that human subjectivity must be removed from its process.

The written record of Greek thought suffered a series of calamities. Early in the 4thC BCE, after Alexander's victories, Alexandria became an important centre of Greek culture. Three centuries later its magnificent library was destroyed and some works were moved to Rome. In the course of 5thC CE, Goth and Vandal sacks of Rome, some of the texts were rescued and taken to Constantinople. As the Christian church expanded into the rest of the Europe, other texts were secreted in various monastic centres. Starting about 1000 CE, many of the Latin and Greek texts, including the Codex Grandeur, the precursor of all Christian bibles, were collected and taken to Northern Ireland. Because of its remote and inhospitable nature, it was assumed the text would be safest there with the "warrior monks", who laboriously produced beautiful copies, which in the 12th and 13thC CE, were returned to Rome under careful guard. In 1452 CE, Mahat the Conqueror sacked the Hogia Sophia in Constantinople and destroyed three quarters of the texts rescued from Alexandria.

Even with all the pressuring measures, it is estimated that 75% of the original texts were lost or destroyed. What remained eventually found its way to the Vatican Library and became the most complete collection of extant western knowledge.

As a result, when significant scientific curiosity once again began to flower, and until libraries were instituted by the Medici's in the 15thC CE, it was to Rome, under the scrutiny of the Church, that the new searches had to come.

The Middle Age approach to knowledge had trained the Western European mind in the basis of rationalism in readiness for the flowering of science. The systems of social organization that had developed to guide both the Roman Empire and the Western Church established a strong sense of the imperative of order -every event had a cause. The belief coming from an insistence on the rationality of God subsumed an implied belief in the order of nature, expressed as the laws of Nature. As one can see in the fallowing, it was the authority of the Church and required behaviour, dependent on faith, that held sway.

In the 13th C CE, St Thomas Aquinas with his advocacy of the importance of the intellect, the senses, and metaphysical revelation in the search for truth exemplified the particular rationalism of deductive scholasticism, inherited by the 16th C CE. The growing ranks of scientists had little question with the supremacy of God. However, they did object to the unguarded rationalism that relied, not on real evidence, but on dialectical reasoning to establish that position. The reaction from science was virtually a revolution – no more dialectic – stay with irreducible stubborn facts.

PROPONENT	MYTH (his truth)
St. Thomas Aquinas 1225 – 1274 CE "Dualist"	The universals (is form) exist in God things and mind. Knowing is to know God – so classify.
William of Ockham 1285-1349 CE "Monist"	Forms of knowledge do not correspond to being. Denied existence of universals except in our minds.
Nicholas Copernicus 1473 – 1543 CE "Monist" (Apriest in Rome)	The truth of universe is absolute and is numbers. The primacy of God is represented by the sun – not man. Universe revolves around sun.
Francis Bacon 1561-1626 CE 'Monist"	Truth can only come from observation rational analysis and verification. Meanings come from sensory experience and experimental evidence not from God.
Galileo Galilei 1564 – 1642CE 'Monist"	Determined by God the cosmos is ordered. Man is separate from it. With no choice it acts through mathematical necessity. It is nothing but math. All sensory experience is illusory, not to be trusted.
Johannes Kepler 1571 – 1630 CE "Monist"	The ultimate expression of God is geometry and musical harmony generated by the orbiting of the planets around the sun locus mundi producing the "music of the spheres".

METAPHOR (how it is done)	RITUAL (the doing)
Truth, affirmation of being as the good comes from both intellect and senses, including revelation.	Dialectical emphasizing derivation of axioms prior to syllogistic reasoning.
Proof of God comes from revelation – is empirical sensory experience not reasoning, which was sophistry.	Open oneself to revelation.
Absolute truth is math. To go against math is to go against God. The planet orbits were circular.	Observations of the heavens wrote his "Discourse on Reason".
Proposed a new priest class to lead society as part of the new Temple of Solomon. (developed into the Royal Society) Dialectic had little value.	His "The New Atlantis" and Descartes discourse spelled the final demise of the dialectic. Disadvantaged merchants and the protestant reformation.
It is clock work mechanical universe set in motion by God. Humanity is separate prom Nature. All humans in common is the ability to understand the cosmos thru math.	Developed the refractive telescope so Venctian merchants could see anything ships first. Found the moons of Jupiter (small around large) Confirmed centrality of sun – the Church livid.
Because everything exists in celestial harmony. it must confirm to math, search of the planets elliptical orbits. The universe is a clock works and not divine.	Using Tycho Brahes (1546 – 1601) 30 years of astronomical observations and math, he looked for the sound – the music of the spheres with no success. Proposed 3 laws of planetary motions.

Following Greece

Bacon and Galileo led the move to reliance on strict empiricism generated by observation and measurement. Reason grew to be based on faith in the demonstrable order of nature. For now, philosophy took a back seat and modern man was born.

From Aquinas to the 17th CE we can see the step-by-step transition from the faith based rationalism of the Middle Ages to tightly held empiricism where reason became based on both faith in the demonstrable order of nature and mathematical discovery of that order. Plato's eternal ideas, "universals", were no longer germane, a change reflected in William of Ockham's 14th C denial of the validity or existence of these universals, except as imagination.

The change in understanding of the structure of the universe is a case in point. Although Plato had felt, based on his philosophical position, that the earth travelled around the sun, the belief that held sway for many centuries was advanced in 150 CE by Ptolemy Claudius Ptolemaeus. His anthropocentric universe had the earth stationary in the centre with the sun, moon and stars revolving around it in circular orbits. The planets were assumed to revolve in small circle, epicycles, the centres of which revolved around the earth in vast circles. Of later observers,not convinced of this classical model, a priest, Nicholas of Cuza in the 15th C CE, postulated that the earth revolved around the sun. The Church was not pleased. Not until the next century Copernicus, and then later, Galileo, with their astronomical observations proved that the entire known universe revolved around the sun. And when, early in the 17th C, Johannes Kepler discovered that the planets' orbits were in fact not circular but elliptical, astronomical science finally had a more or less correct metaphor for its limited understanding of the universe – an example of the evolution of both mythology and metaphor.

Coupled with the growing insistence on the supremacy of mathematical order in nature was the development of advanced technology for measurement and thereby evidential verification of that order. Aristotle's idea of "Quality"was no longer king. Until the appearance of Relativity, Quantity, numbers, was it again – they were real. Introspective thought could not be measured and one could no longer believe without proof. It is an attitude that still rules today even though ,in many cases, that proof is quite abstract and often largely theoretical.

To say more about this transition from the Middle Age emphasis on non-demonstrative metaphors and abstract dialectical reasoning,we begin to see a return to classical metaphors. The Pythagorean supremacy of numbers and there from mathematics, beginning with Nicholas of Cuza, again began to hold sway. The Aristotelian importance of non-theoretical observation of the physical world was again recognized. Although Aquinus and Ockham had claimed validity of sensory experience, there was a gradual removal from this basis, partly

Following Greece 25

due to the fear that emotions could adulterate that experience. Galileo claimed that all sensory experience was illusory, not to be trusted or used, an attitude maintained well into the future. One has to ask how this position was reconciled with the fact that the senses, particularly vision, were the primary providers of the 'hard' data they had access to.

With Galileo, Europe entered the modern age. Though he placed his ultimate faith in the inevitability of God's cosmos, it was his mathematical lens through which he explained the universe. He was concerned with the nature of motion. He approached that investigation by first imaginatively constructing an ideal situation in which a moving object was launched above a horizontal plane, subject to no hindrance.

Obviously in the real world, hindrance would exist – call them the facts. His observations would then consist of noting the relationship that existed between his imagined reality as a baseline, and the effect of those facts. Combining creative imagination and hard observational data enabled him to explain the true nature of movement – Kinamatics – a genesis of modern physics.

This ritual produced a proposition that the laws of mechanics are equally valid in all frames of reference that move uniformly in relation to each other – his metaphor of a belief that we shall see translated again by both Newton and Einstein.

The belief in the primacy of mathematics held by

Nicholas of Cuza, Copernicus and Galileo was made even firmer by Rene Descartes. Although there was consistency in that fundamental belief, their individual metaphorical interpretations varied, each of their rituals purported to enact the myth, but never exactly, thus gradually altering the myth – this has been the process from the beginning.

Descartes, who maintained "I think, therefore I am"held that quantitative properties perceived with reason, through mathematics and measurement, provided the only true outer reality of the physical world. This position echoed Plato's belief which had contended that mathematics and numerical ratio produce more certainty than evidence from the senses. Descarte's version held that the qualitative, discerned through sense perception, did not describe outer reality. We must not assume that this virtual agreement made these ideas true. It was just another indication that old myths never die. They are just inflected in different ways.

Descartes has been termed the father of analytical geometry. Using his cartesian co-ordinates and his invention of graphing, he was physically able to demonstrate his truth – numbers were given form. His graphing was based on plain geometry – straight lines. With reference to a then apriori body of mathematical knowledge defining a straight line as passing through three points, he extrapolated the possibility of a fourth point, or more, that extended that line.

PROPONENT	MYTH (his truth)
Rene Descartes 1596 – 1650 CE "Dualist" (from Galileo)	Truth exists and is only understood thru Gods creation – mathematics. The universe was a great machine. A perfect entity must originate in God – the perfect entity.
Isaac Newton 1642 – 1727 CE "Dualist"	Truth exists thru numbers in mathematics. Time was absolute. The everlasting mechanistic universe, set in motion by God worked like a giant clock.
James Maxwell 1831 – 1879 CE "Dualist"	Thurth exists and it was all found in Newton's "Philosophia and Natural's Principia Mathematica".
Albert Einstein 1879 – 1955 CE "Dualist"	Truth was implied in his belief in absolute universe everlasting to everlasting within which everything is relative except the speed of light the only invariant. He believed a doctrine of a Grand Unified Field Theory (GUT) must exist – leading to the ongoing debate between the Relativists the Quantomists and the super stingers, all contingent Godel's Incompleteness theory.

MTAPHOR (how its done)	RITUAL (the doing)
Agreed with Galileo that man is separate form all Nature. Man's mind is separate from his body and senses, which can not be trusted. Mind was unchanging – the only link with God using rational analysis.	Invented graphing and Cartesian co – ordinates, making possible extrapolation to predict truth – "but don't exceed the date base". Father of analytical geometry. Published "Discourse on Method".
Man is separate from all Nature senses not to be trusted. Along with Leibriz, he invented calculus to give God's laws on how the clock worked. His absolute metaphor is his Philosophia Naturalis, Principia Mathematica.	Used the dialectic and the calculus to understand the mechanistic universe. Melded observation, physics and calculus to formulate his 3 laws of Motion and the law of Universal Gravitation.
Postulated action between entities at a distance thru force measurable in accord with Newton's Law of Universal Gravitation. Propose the speed of light – "C"and light as a wave phenomenon.	Developed electromagnetic theory for eguations describing properties of light and electromagnetic spectrum. Using Newton's Law he demonstrated force field and lines of force with magnet and iron filings.
Mass is equivalent to energy ($E = Mc^2$) Qualitative is equivalent to Quantitative. Together they create and wrap space/time and dictate behavior of Matter. Motions produce by Gravity and Inertia are equivalent and can only be judged with respect to a frame of reference – is the speed of light. He advanced the particulate theory of the nature of light.	Using mathematics, he developed the special and General Theories of Relativity in response to Houbble's "red shift"finding that demonstrated an expending universe he fudget his math with his Gravitational constant to reconcile with his believe in the absolute universe until that math was disproved by Le Maitre. He failed to create a Grad Unified field theory to link Quantum and Relativity.

Following Greece

This may or may not have been true, and in response to this lack of certainty, science later cautioned against "exceeding the data base"–the three points. Obviously the limited value of his ritual became evident once Newton co-developed solid geometry and calculus.

In his publication of "Discourse on Method"- Descartes attempted to convince philosophers that, using reason as the only path to knowledge, their process must parallel science's use of it's Demonstrative Methodology. He also held that modern professionalism should involve the training of minds to conform to the methodology, a limiting axiom still practised today.

Another of his proposals had early ramifications for the future understanding of the universe. He argued that the mere separation of bodies by distance proved the existence of a medium between them – space had substance. This idea led in the following century to the proposal that light, then seen as consisting of waves must, like water, have a medium to support them. The substance was called "ether"and opened the doors for Faraday's and Maxwell's theories of electric and magnetic forces.

The debate over the validity of sensory experience swung back and forth. The British Empiricist School of Philosophers, Hobbes and Locke in the 17th C CE Hume and Berkley in the 18th CE, were firmly resolved on the centrality of sensory information in that all meaning came from the senses.

Although not recognized at the time, this position was consistent with science's reliance on observation in its method – obviously one observes with one's senses. The dialectic, except for its use in the analysis of the results of those observations, had no fundamental value for either camp.

Another aspect of contention played its part in the debate. Since the middle ages the Roman Church had supported reason as the primary pillar of faith and thereby guided certain of the scientific community. The more recent Protestant position placed its emphasis on unreasoned faith. This divergence resulted in significant political alliances – one had to know where the power lived.

In the midst of all these dichotomies was born a scientist that provided the next step into what today we consider to be modern sciences. In fact many of his positions still underlie the general population's impression of how the universe works. Isaac Newton, born in 1642, shortly before the death of Descartes, continued using the world of numbers and mathematics to explain God's creation. In his clockwork universe time was absolute, the only invariant, and which provided a constant frame of reference for his theories – not to be challenged for another two hundred years by Relativity.

Actually, following along the path indicated by Galileo and Descartes, essentially paraphrasing Galileo, Newton developed his own relativity principle which held that mechanical laws which were valid in one place were equally valid in any other place that moved uniformly, relative to the first. This understanding led to his laws of motion which stated that a) an object continues in its state of rest or motion in a straight line unless acted on by a force, b) Force=mass x acceleration and c), every action is accompanied by an equal and opposite reaction.

He also recognized gravity, which he described as "action at a distance"and proportional to the mass of the object in question. He understood that what caused his famous apple to fall and also kept the moon and the planets in orbit around the sun. He could describe this but could not yet explain it.

It is important to note that Newtonian physics made the space program possible three hundred years later. Knowing through astronomical observations the motions of the earth and moon relative to each other, it was possible to compute a straight line cartesian path for a spacecraft to intersect with the moon or any other planetary body. Newton's assumption of the mechanistic universe as a great clockwork allowed full prediction because of course it was all predetermined – from the beginning of time. Knowing the position and momentum of an object in space, straight line and solid geometry would supposedly allow future predictions of its location. The fundamental lesson of Newtonian physics rested in the belief that the universe was governed by laws that were manifestations of God's perfection and were therefore open to rational understanding.

The significance of Newton's propositions must also be seen with the understanding that they only work for the large scale and for low energy fields, and not in the realm of the very small, sub-atomic world. Also with regard for his claim for predictability, Quantum Mechanics, in its belief in the fundamental atomic nature of creation, later claimed that one could only determine probability but never predict future events with certainty.

In 1687 at the age of forty five, Newton published his "Philosophiae Naturalis, Principia, Mathematica". Incorporating the substance of his "truth"it became the bible of science for more than two centuries. Included in his theses was the contention that light had a particulate form, what we today call photons. Christian Huygens, a contemporary, had developed a wave theory of light, quite opposed to Newton's corpuscular theory. This debate was to continue, reaching a dramatic hiatus with Einstein at the 1927 Copenhagen Conference – to be discussed later on.

The other ongoing debate between rationalism and empiricism over their relative legitimacy in the gaining of knowledge was given new energy through the voice of the 18th century philosopher Immanuel Kant. He recognized the two elements that must be considered: the external information, known through sensory experience and the internal "form"– conditions of the observers mind. (the red glasses he may be wearing) and absolute eternal time, space and causality, not experiencable as an event. He explained sensory experience in terms of mental conceptions, equivalent to but not identical with the object of experience. Reason, without the necessary sensory material to process, could not deal with the big questions like proving God's existence or whether or not the world had a beginning. Contrary to Plato, he felt that ideas being free of experience could not be truths nor could they be demonstrable. Therefore he completely rejected the dialectic and the use of metaphor. His "3 Critiques of Reason,"useful for a greater understanding of human cognition, held that knowledge was gained using the categories of logic and through space and time. His thoughts were profound and served to unite the rationalists and empiricists but failed to answer how to put those goals into action. The limitations his positions imposed on the potential for knowing, although may be appropriate for the time, created, relative to truth, an illusion. "Apriori knowledge existed".

Light was mentioned earlier and we will get back to it, but the evolution of the understanding of colour is an example of the illusory nature of some positions held by science. Newton, using the model suggested by the performance of a prism said that light existed in discrete bands of colour with clear separation between each part of the colour spectrum we often describe as "ROYGBIV". A century later the philosopher J. W. Goethe suggested that there must be colours between the colours. In fact a virtually infinite variety should be possible. Of course artists already knew this but it wasn't until demonstrated by the Ostwald colour wheel system and by the 20th Century development of four colour separation and the variety it allowed, that the truth was verified. Eighteenth and 19th century physicists felt that if light was a wave phenomenon, by then generally accepted and proven, then there must be a medium to support the waves, something to wave. As a result it was held that the entire universe was permeated by 'ether' an invisible tasteless, colourless substance that allowed the propagation of light waves. James Clark Maxwell, whose "truth"was completely stated in Newton's "Principia"disagreed however with that position on the nature of light. He believed that light consisted of electromagnetic waves. His field equations demonstrated that light propagation could also be envisioned as energy propagation in empty space (in vacuo). The ether theory was eventually disproved experimentally by Michelson-Morley, finally put to rest by Einstein's denial of its existence. Maxwell and Faraday's demonstration that a magnet creates certain properties in space led to Einstein's conclusion that celestial objects affect the gravitational field around them.

Introduction to Relativity etc.

As we move into the world of Einstein and beyond, it is useful to compare the essential characteristics of Newtonian, Relativity, Quantum and Chaos Theories on the level of their equivalence and their differences. Each of them expected to develop the theory of everything. First, to get an initial idea of the differences between the first three, consider them in terms of a metaphorical model of a billiard table. To begin, both Newtonian and Relativity use the same table. For Newtonian the table is fixed and flat. Balls travel in virtually straight lines as we would normally expect and then bounce off the edges in a predictable manner.

In General Relativity, the balls are different sizes. As they move over the apparently same table the edges and supposedly flat surface change and warp, thereby influencing the ball's movement. The faster or slower the balls move the more or less is the influence. The harder the cue ball is hit, the more we are into Relativity. As the hit becomes softer, we approach Newtonian.

Now, for the Quantum table, things are completely different – it is a virtual table. You are blind

– folded, you don't know how big the table is, how many balls there are, nor how big they are. There is a probability that one of the balls might eventually find a pocket – but with no certainty. Playing pool on the Quantum table is like trying to pin the tail on the donkey with no outside guidance. And for Chaos, there is an infinite number of tables and balls and all are connected.

There is an equivalence between all four of the theories, but only partial. Note in the following chart that only three have a geometrical representation and only three have a tensor or mathematical manifold. This lack of complete coherence challenges our ability to relate them. (Note also that a manifold can be likened to a transmission that connects auto elements).

FOUR METAPHORS OF EQUIVALENCE				
	Newtonian	Relativity	Quantum	Chaos
Geometric Tensor	2D Circle 3D Ellipse	4D Hyperbola	None Possible	Attractor (Point, Period or Strange)
Math Tensor	linear Algebra & Geometry	linear Matrix Algebra	linear Quantum Mechanics (Stochastic)	non-linear Fractal, Lyapunov, Hamming

Note: Tensors were seen as math arrays ,(matrices) also as operators (circles, ellipses and hyperbolas) in Relativity theory, that map vectors onto co-vectors. Relativity's geometric tensors(vs) Quantum's math tensors.

Introduction to Relativity etc.

The Manifold Approach

Newton's mechanistic universe, set in motion by God, was closed and completely deterministic. Einstein also believed in an absolute universe "everlasting to everlasting" but in which everything was constantly changing, relative to itself – all possibilities within absolute boundaries.

He called the approach "manifold". Being inclusive it might link disparate elements. With no dogma of separation, knowing there always are new data it assumes a potential discovery of a unity of all theories. It is founded on three definitions:

1. There is no origin point everything is relative.
2. There are sets that can be put into a series of one to one invertable transformations e.g. Lorentz.
3. There are spaces of arbitrary dimension that locally appear cartesian – i.e., new data.

On these bases, new metaphors can surface as one steps outside events, thus changing perspective. For instance, exploring relationships and equivalences between general Relativity (4D curved space) the local effects of Special Relativity (3D/4D) flat space, and the weak field limit of Newton (2D/3D flat space) is a manifold approach which would work to unify or reconcile different beliefs.

In order to achieve unity, or even a link between different theories, their basic beliefs or myths and their metaphors must be relatable. Also for the manifold approach to work, each theory must have a visualizable component - geometry or mathematical language. For example in attempting to link Relativity Quantum and Chaos you will see in the previous chart the inherent impediments:

- There is no geometry to link Relativity to Quantum, which has none.
- There is no linear math to link Relativity or Quantum to Chaos, which has none.
- Relativity says that the macro high energy universe is geometrical and can prove it.
- Quantum says that the micro-low energy universe is not geometrical and can prove it. NB- both use math that is philosophically linked but not scientifically provable.
- Quantum says the universe is governed by the laws of probability.
- Einstein said its causal and "God doesn't play dice with the universe"
- Chaos has a qualitative, non-linear mythology – Quantum, qualitative and linear.

Even with all of these reservations, science is still attempting to find links. Newtonian was shown its limitations by Relativity, Quantum and Relativity arrived at a virtual stand-off and Chaos may be the new hope in the search for the "theory of everything" By being non-linear, it may have the edge – my bias.

Albert Einstein

What a shock it must have been for the world of science when Einstein and Relativity turned up on the scene. For more than two hundred years, the universe had been understood in terms of Newton's clockwork machine, operating deterministically in accord with his Laws of Mechanics and Euclidean Geometry, straight lines and circles. The only absolute was time – the everlasting element. Then along comes this German mathematician claiming that time was only a fourth dimension, a perception, and that the only invariant in a universe that was constantly changing was the speed of light. If you had been a Newtonian what would you have thought?

Like Plato and Newton, Einstein's Universe, everlasting to everlasting, was absolute, but only in terms of its extent, its constituent materials and the Laws of Physics. Within these boundaries, everything was relative, a restless universe in which motion could only be considered in terms of being relative to some other frame of reference, also moving. No longer was it appropriate for science to look for an absolutely stationary frame of reference. The new mythology stood that idea on its ear.

Born in 1879, two centuries after Newton's publication of his "Princip", Einstein showed an early interest in science and mathematics. He found employment in a Swiss patent office where part of his duty was to check applications for violations of the 2nd Law of Thermodynamics., i.e., that all closed systems move toward disorder (entropy) at the expense of order (neg - entropy). Among the ideas that challenged his curiosity was the 1881 Michelson Morley experiment that demonstrated that the speed of light, accurately determined in 1849 by Maxwell as 186,284 mi/sec was not affected by motion of either its source or its receiver. Einstein reasoned that, this being the case, its speed would be constant regardless of the motion of the earth or any star or planet moving in the universe. In 1905 he generalized this understanding into his Special Theory of Relativity which stated that all the laws of nature are the same for all systems that move uniformly, relative to each other. More than just mechanical laws referred to by Galileo and Newton, his theory included light and electromagnetic phenomena. He believed that to explain the universe rationally, there must be a confidence in the harmony and order of nature. This core belief led him eventually to utter his famous line, "God doesn't play dice the the universe. There are no accidents". The laws of Nature, understood or not, always apply.

He asserted that to understand these laws in terms that are consistent throughout the universe one must view measurement of time and distance as variable quantities. At what we think of as normal speeds, virtually stationary compared to the speed of light, it is quite appropriate to add two speeds. For instance, you are on a train moving at fifty miles/hour and you are walking down the aisle in the direction of travel at three miles/hour. Your actual velocity relative to the track below, would be fifty-three miles/ hour. Walking towards the rear, it would be forty seven miles/hour. Normal speeds have virtually no effect on time or space. However, as velocities move toward the speed of light this is no longer true.

In 1893, H.A. Lorentz, a Dutch physicist, working on a somewhat related problem, developed with the help of the French mathematician Poincare, a set of calculations that allowed Einstein to advance the verification of the implications of Relativity. As a result he added the axiom that "the laws of nature preserve their uniformity in all systems when related by the Lorentz Transformations."

Two of those transformations from the observer's viewpoint are as follows:

Note – it is not possible to have a minus quantity and the speed of light can never be exceeded.

$$M_f = \frac{M_o}{\sqrt{1-\frac{v^2}{c^2}}} \quad \left\{ \begin{array}{l} \text{as } v \text{ approaches } c \\ \text{mass} \approx \infty \end{array} \right.$$

$$t_f = \frac{\sqrt{1-\frac{v^2}{c^2}}}{t_o} \quad \left\{ \begin{array}{l} \text{as } v \text{ approaches } c \\ \text{time} \approx 0 \end{array} \right.$$

M = mass
t = time
o = initial
f = final
v = velocity
c = speed of light

Albert Einstein

In other versions of these transformations we find more surprising results as the speed of light is approached. For instance:

- Lengths contract, becoming zero at V=C
- Clocks slow, stopping altogether at V=C
- A rocket launched at a velocity of 0.75C from a spaceship moving away from the earth at a velocity of 0.75C would have a total velocity, not a sum of the two, but of 0.96C with respect to the earth.

Remember, space and time are virtually unaffected by motion at low velocities. All though completely consistent with Einstein's theory, for the classical expectations to be so challenged must have been a bitter pill for his peers to swallow. When his Special Theory was expanded to remove the need for "uniform motion" science had an even tougher job catching up.

Up to the 20th Century physics was still of the general opinion that light was emitted in the form of waves. In 1900, Max Planck theorized that bodies emitting radiant energy did so, not in a continuous stream, but in discontinuous bits which he termed "quanta". Although not explained, his mathematics showed a constant relationship between the amount of energy emitted and the frequency of the radiation. Einstein, almost alone amongst his contemporaries, appreciated the theorie's significance and led him to propose that all light consisted of discrete particles to be called "photons". Five years later he proposed an explanation of what was known as the photo-electric effect. Physics had been unable to explain, thru the classic wave theory why, when light fell on a metal plate a shower of electrons was ejected. So when a theory can't explain phenomena, change the theory. Einstein's position that light was not a continuous stream but his discrete photon particles was seriously challenged later, but for his photoelectric effect he was awarded his only Nobel Prize.

Although it wasn't published until 1915, along with his General Relativity theory, he early on understood the equivalence of energy and mass, expressed in the famous short form equation E=MC2. In its complete form the equation, because it accounted for the influence of velocity, stood not for equivalence but, in the causal universe only, for identicality.

$$E = \frac{Mc^2}{\sqrt{1 - \frac{v^2}{c^2}}}$$

You can see that low velocities will have virtually no effect on the calculation, but as velocity approaches the speed of light, the denominator reduces, thus resulting in an accelerating increase in both energy and mass.

In another area, Einstein had never been happy with the current position on gravity. You will recall that Newton with his "action at a distance"theory proposed in his Law of Inertia that the forces of inertia and gravity balance each other. Gravity force was exerted sufficient to over come the inertia of an object. Contrary to Newton and his contention that things could be at rest, Einstein knew that everything was constantly moving. Imagining how 2 people inside a closed container would experience free fall, accelerating upwards or attached to a spinning merry-go-round, he developed in 1912, his Principle of Equivalence of Gravitation and Inertia. Simply stated there is no way to distinguish the motion produced by inertial forces (i.e. acceleration, centrifugal force, recoil, etc.) from motion produced in a gravitational field. For instance, the effect produced in an aircraft pulling out of a dive is the same as that produced in a high speed steeply banked turn. There is no up or down in what he termed a gravitational field, which is created by context and circumstance.

It wasn't a matter of force at all. Force he considered to be an illusory idea inherited from the old mechanistic universe model. As a magnet and iron filings demonstrate a magnetic field, Einstein's Law of Gravitation saw gravitation fields as describing the field properties including geometry, of the space-time continuum as it bends the path of objects – planets for example – moving through it.

Celestial objects individually determine the properties of the space around them. This principle became the cornerstone of General Relativity which was no longer tied to uniform motion. Motion, both uniform and non-uniform, can only be judged with respect to some system of reference – there was no absolute canonical, inertial frame of reference for motions – all are relative.

Space-time should be seen as space with time. The three dimensions of space, plus time, cannot be separated. Being four dimensional ourselves we cannot see the geometry of that indivisible space-time continuum. We can however try an analogy. The continuum could be likened to the surface of a hilly landscape. Relative to a flat plain, hills result from an excess of matter, valleys from a lack of it. These characteristics could be said to result in the curves and undulations of its surface. Ina parallel way, bodies of matter in space cause bumps or undulations in the space-time continuum – the larger the body (a mountain), the larger the bump in the gravitational field. In the remoter parts of the universe, far from bodies of significance, the continuum could be likened to a flat prairie, without curves and undulations. An object moving through space would take the shortest path from A to B as it responded to the diverse spatial neighbourhood architecture through which it was passing – the warped space-time continuum created by gravity. In Relativity Theory, every path through space-time is based on the geometry of an hyperbola and forms geodesic curves, e.g., an arc diameter of approximately.2 light years, the smallest arc segment of which could appear to be a straight line, an illusion that could suggest a cartesian geometry. A new non-euclidean geometry and a new mathematical description was required. It was another step along the path towards the goal of a "universal theory of everything, "even though that understanding was limited by that very same geometry and mathematics.

In 1915, the General Theory of Relativity was published. It stated that "the laws of nature are the same for all systems regardless of their state of motion". We need to bear in mind that Einstein's universe was huge, but finite. The Theory referred to a closed system. It also gave a greater weight to the idea of the equivalence of mass and energy. Light, being energy, had mass. Its path would therefore be affected as it moved through a gravitational field. It would cause the lights path to bend. He called it the Gravitational Lense Effect. His theoretical calculations, as to what deviation would be, formed the first of three major predictions that General Relativity would generate. This first prediction was verified by Sit Arthur Eddington, who in his observations of the sun found a deviation of a stars light, passing close to the sun, to be 1.64 seconds of arc, 6% off the original calculation.

Next, if one were using Newtonian laws, one would expect planetary orbits to be constant and fixed. However, these laws could not account for the gradually advancing elliptical orbit of Mercury. Even though this perturbation was only 0.43 seconds of arc, per century, Einstein was able to satisfactorily explain this anomaly using his laws of gravity. It was also confirmed by Eddington.

The third prediction dealt with what he termed "time dilation". Having shown how the properties of space are affected by gravitational fields he went on to show how time intervals also vary under these influences. A clock on the sun would run at a slightly slower rate than one on earth. A twin on an extended space voyage would experience life as usual, while his brother on earth would see him living in slow motion and when returning to earth would be younger. Newton was wrong. There is no absolute single time that flows equally for all observers. "Now" could only be a relative term. Time dilation was proved in 1972 flying atomic clocks around the world.

But not everything was roses. Einstein realized an anomaly in his own theory. General Relativity unexpectedly implied an expanding universe. Being still rooted in the belief in a finite one, he could not accept this implication in his theory, so he concocted a 'gravitational constant", a fudge factor that would allow alignment of mathematical theory with belief, a case of Ritual enacting into Myth. Meanwhile another surprise was surfacing.

PROPONENTS	MYTH (his truth)
Quantum Neils Bohr 1885-1962 CE (Quantum's Father)	Bohr believed there had to be a system that worked for the micro universe, which Newton and Relativity didn't do.
Erwin Schrodinger 1887-1961 CE	Quantum a field of all possibilities, but no certainty. Not this or that but this and that – with probability.
Werner Heisenberg 1901-1976 CE	One must choose what one wants to know and observe. At the sub atomic level, it is not possible to observe reality w/o changing it we are part of the nature we study.
	The death of local causes.
CHAOS (fathers) Edward Lorenz James Yorke Mitchell Feigenbaom Benoit Mandelbrot	Chaos means hidden order and concerns it self with non liner complex dynamical systems in space and time that are deterministic in the short term and poses the properties of emergence, irreducibility and an inordinate sensitivity to initial conditions. (Initial Condition)
VEDIC Science Vasistha (Sage)	The reality of creation is that it is totally unified, undifferentiated and bounded.

METAPHOR (how its done)	RITUAL (the doing)
With complementary Bohr showed how light exhibits either wave like or particle like behaviour depending of how we choose to interact with it – ie which experiment.	Used scientific method on the micro of relative value even with limited observational evidence.
Shrodngers "cat in the box" experiment – opening the box resulted in the collapse of the way function (alive or dead).	Mathematically predict the probability of an event taking place and develop statistics on the behaviour of population growth.
Heisenberg uncertainty principle = one can't know both position and momentum of an election at the same time. The more we know of one the less of the other.	Relativity's matrix algebra didn't demonstrate the commutative principle for Quantum i.e. A=B=B=A
Superlumianl communication	Bell's Theorem
Attractors exist in the phase space and represent all the possible states of the complex dynamical system. I.e., all possibilities of what can and what can't happen. Because initial conditions never repeat exactly, it is the river that you can never step in twice. (Attractor)	Iteration of the system may result in emergence of islands of stability (i.e., emergent chaos) and thereby evolution of the metaphor and additions to the myth. (Interation)
The appearance of a diversified material universe is illusion Maya – a creation of mind.	Knowing this truth, we act as if material creation is real and, playing the game, do what needs doing.

Albert Einstein

The astronomer Edwin Hubble, working at the Mt. Wilson observatory, noticed a shift to red in the wave length of light coming from distant stars. He reasoned that the increase in wave length towards the red end of the colour spectrum could only result from the source of light receding, thus increasing its wave length. Humison, his assistant, like Tycho Brahe for Kepler, carried out exhaustive analysis of the spectrographic data and verified Hubble's conclusion that the universe was actually expanding. It was not absolute. Einstein remained adamant only whilst the mathematician Abbé Le Maitre, using Einstein's math, proved that Hubble and Humison were correct. Einstein capitulated and threw out his "gravitational constant" Hubble got the Nobel Prize (Humison ignored).

While all this was going on, the new Quantum theory was developing and Einstein was to be involved in a major contra temps over the true nature of light - wave or particle- at the 1927 Copenhagen Conference.

Even though the Quantumists were essentially children of Relativity, the two positions were completely non-transferable. This becomes quite obvious when we compare their characteristics.

RELATIVITY	QUANTUM
Invariance	Variance
Is or isn't	Can be both
Deterministic	Non-deterministic

Demonstrable re the macro universe	Demonstrable re the micro universe
Geometrical	Statistical
Geometry matrix	Probability matrix
Certainty	No certainty
Visualizable	Non-visualizable
God doesn't play dice w/ universe	Does play dice and sometimes cheats

The two positions spoke completely different languages. It is no wonder that their communications were so unresolved. One saw the universe through Hubble's telescope, the other through an electron microscope.

Quantum arose out of necessity. Relativity couldn't answer many of the questions about the atomic and sub-atomic world. At the same time, quantum could address 3 of the fundamental forces, but not gravity which is a geometric, visual tensor. It had carved out a new world for itself, a world not readily accepted or fully understood by it proponents.

Up to this point in our survey, the particular myths have been an expression of the individual proponent's belief systems, even though also a development of previously held positions.

In the cases of Quantum and Chaos, as the following table demonstrates, there is no single author of the belief system that underlies either theory. The authorship is best described as a collective that through its parts, attempts to describe the whole, which in Quantum's case was coordinated by Neils Bohr. In both cases the multiplicity of beliefs leads to a complexity that discourages our ability to settle on a simple coherent description.

Quantum Theory

The term "quanta" was coined by Max Planck in 1900, referring to the discontinuous discrete packets of energy emitted or absorbed by a body or an atom. Prior to the introduction of Quantum theory, Ernest Rutherford, in 1911, had described the the atom's form – a planetary model still in vogue today. Like bodies in our solar system, electrons were seen as discrete particles that orbited around the nucleus at discrete distances from it.

Carrying on from Planck's proposal, the quantum model for the atom proposed by Neils Bohr in 1913 was a markedly different metaphor – a non-three dimensional model. He proposed energy shells corresponding to Rutherford's orbits. Bohr described the shells as spheres of probability as to where the electrons were to be found – but no certainty of exactly where. Each shell represented discrete quanta of energy with nothing existing between the shells. Like the understanding of the atom before 1900, a quanta, or packet, was now described by Bohr to be the smallest indivisible quantity of energy a system could present.

Here we go! After years of focus on the infinitely large, onto the stage appears a movement only concerned with the assumed infinitely small. Some ten years after Bohr's proposal, Wolfgang Pauli's "exclusion principle"stated that no two electrons could occupy the same position in space-time and suggested that electrons existed metaphorically as particles within the probability shells, or energy levels, surrounding the nucleus. None of these electrons in an atom could be exactly alike. Depending on the particular number of electrons, the required number of shells would exist around the initial shell at discrete distances from each other. As the atom gave off or took in energy, i.e., lost or gained an electron, the quantum would jump from one shell to another, giving off its discrete quantity of energy.

To try to visualize what is virtually unvisualizable, think of these shells as looking like a cloud – like substance, separated from each other by a much less dense smoke-like quality.

Quantum Theory is rooted, not in certainty, but in probability- all possibilities exist. Not this OR that, but this AND that. Erwin Schrödinger proposed a situation which became a classic dilemma speaking to the theory. A cat is placed in a box that contains a device that can release a deadly gas should a random event trigger it. The box is closed. Classical physics would have it that the cat is alive or dead. Quantum says that before the box is opened the cat is in a kind of limbo represented by a wave-function containing the possibilities that the cat is alive and dead. When we look into the box one possibility actualizes and the other has vanished. This is described as the collapse of the wave function. The outcome isn't there until we observe it. Before that it is only a wave function. Each time the box is opened, the quantum wave function collapses from alive AND dead to alive OR dead.

Quantum means "discrete wave packet." If it could be displayed on an oscilloscope as a quantum of energy, it would look like this.

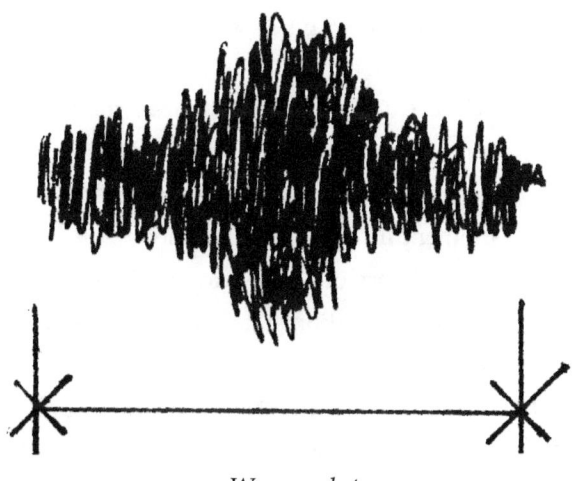

Wave packet

In Relativity, the history of events through space-time was described by H. Minkowski, Einstein's mathematics teacher, as a world line- a relationship to the "real world". In Quantum a wave packet is analogous to an "event"in Relativity. The history of the wave packets in space-time is known as a "quantum linear super position state" – analogous to a world line. Though both theories rely on differential math, unlike Relativity's certainty, analysis of this quanta-history can only indicate real world probabilities. Of the many formulae, Quantum Mechanics presents, for those conversant with mathematics, the following series ties the quantum state into the "real world".

ψ	=	Quantum state or wave function of a particle
$\psi(x)$	=	For each position in that quantum state there is a specific value – the quantum complex amplitude donsity
$\|\psi(x)\|^2$	=	A wave packet – the squared medulous of a quantum complex amplitude density, which is not exact and gives the particles position and momentum over a small region
$\Sigma\|\psi(x)\|^2$	=	Total history of all the wave packets (i.e., events), the super position state giving all possibilities.

In terms of the "cat in the box" model, each wave packet is the cat box. Not necessary to look everywhere, the box contains all possibilities for the cat, including location – but only before looking.

The old see-saw debate over the true nature of light – particle or wave – became central in the development of Quantum Theory. Recall that Newton said it was particulate, and his colleague, Christian Huygens said, "No, it was a wave phenomenon. "In the early 18th century, Thomas Young, using a version of the double slit experiment, proved Huygens correct, later to be agreed with by Maxwell. Then 1900 saw Max Planck's proposal that light consisted of small packets of energy he called "quanta". Einstein agreed – discrete particles which came to be known as "photons"– a position that seemed to be questioned by the double slit experiment.

So why do I go through this list? – mainly to illustrate the reality that, at any time, what a scientist is fervently supporting right then, is only a temporary metaphor, based on Fisher Information and not an absolute truth. As an analogy, I recall my philosophy professor defining this subject as "what philosophers are talking about today".

Richard Feynman suggested that by understanding the double slit experiment, all Quantum mechanics could be gleaned. The set-up for this experiment has a beam of light directed at a piece of cardboard containing two vertical slits, and set in front of a photographic plate. When one slit is open the plate records as expected, a vertical bar of light. If light were particulate it would create this coherent image. However, when both slits are open, what appears is not just two bars but a series of light and dark bars, a pattern resulting from the interference between the patterns produced by the two light "waves"passing through the slits. You could think of two wave patterns approaching each other on the surface of a pond. As they collide, two coinciding peaks will reinforce each other, as will two troughs.

When a wave and trough coincide, they will cancel each other. Hence, in the experiment we see an alternating pattern of light and dark – not the result predicted by those who believed light consisted of a continuous stream of particles, or photons. Combine these results with the photo electric effect, the dilemma was eventually handled by proposing that light had both wave-like and particle-like properties, depending on the experiment used – and humorously called a "wavicle".

The proposer of the label, Prince Louis de Broglia, in 1925 suggested that, in the interaction between matter and radiation, electrons, which up to then had been conceived of as particles, could in fact be best understood as systems of waves. The behaviour of electrons, having no definite position in space or specific volume, seemed to be too complex to be a material particle. Stimulated by de Broglie's proposition, Erwin Schrodinger developed a system of mathematics called "wave mechanics", that attributed specific wave functions to protons and electrons. In fact, he reinterpreted Neils Bohr's energy shells of electrons to actually be standing wave patterns, specific to the number of electrons in each shell. Without knowing what was really waving, he refered to whatever it was as apsi () or wave function. His equation governed the shape and evolution of these as probability waves. Shortly after this, two American scientists experimentally verified that electrons do exhibit wave characteristics.

Not everyone was captivated by the Quantum propositions. It was held by the logical empiricist camp that it was naive to speculate on the true nature of anything. Science could do little more than report its observations. If in using two different instruments, light was seen as both wave and particle, then both results must be accepted in describing reality. It became evident that even more refined instrumentation was not going to lead to truth. The capricious microcosm resisted exposing all of its secrets, if in fact they had the appropriate metaphors. In fact it was a world plagued with uncertainty and the debate persisted.

A significant idea was put on the table in 1927 when Wiener Heisenberg proposed his Principle of Uncertainty which asserted that it was impossible, with any of the theories then known to science, to determine both the position and momentum of an electron at the same time. The very act of observing one changed the other. Quantum's emphasis on probability shook the pillars of the old science - both causality and determinism. The stage was set fro the historic battle royal between the Relativists and the Quantumists at the 1927 Copenhagen Conference.

At the time of that conference, while at least 90% of Europe held to the conservative positions of Newton and Darwin, there was growing interest in the new theories of Relativity and Quantum. The Conference was to break new ground in its attempt to reconcile these new theories. Berlin, the centre of all of Einstein's Relativists, was scientifically stable but socially unstable. The concurrent rise of ideas of social relativism honoured a non-absolute context – an avant garde freedom that fostered socialism, liberalism, economic expansion, and moral decadence, all encouraged by the cultural implications of Relativity. The old order was breaking down.

On the other hand, Copenhagen had always been more socially liberal than Berlin and Germany.

Social and financial stability were almost a tradition for Denmark. Science was a different matter there. Even though Neils Bohr and the Quantumists were mostly children of Relativity, their proposal of Quantum Theory represented a radical departure from their "parents"position and was deemed by them as being scientifically unstable. We recall Einstein's "god doesn't play dice "response to the quantum emphasis on probability. The conference was a magnificent clash of metaphors one in which the two protagonists were loth to speak to each other. Even if they wanted to, they hadn't a common language.

The Quantumists had their own metaphors of the micro universe to reconcile. The experience of the double slit experiment only added confusion to the debate:

- Schrödinger (like Huygens) said "Its a wave if

you don't look", determinism (cause and effect) applies
- Heisenberg (like Newton) said "Its a particle "If you do look, uncertainty applies. de Broglie of course said "Its both"

Bohr, the referee said "These mutually exclusive characteristics are complimentary aspects of light but both can't exist at the same time. His concept of complementarity included what bordered on subjectivity, a no-no for logical empiricists in its suggestion that the common denominator in the correlation of all the experiences of the physical world is not external reality, but the "I" that experiences and interacts with that reality. It implies the need for a greater understanding of the nature of observation. I have to ask –do all observers translate what they see through their own metaphors and if so what chance is there for coherence and that much sought after "truth".

The unfolding of Quantum Mechanics was (and still is) a drama of high suspense. After a conversation with Bohr in 1927, Heisenberg repeatedly asked himself "Can nature possibly be as absurd as it seemed to us in these atomic experiments?" How could physicists predict something, mathematically describe it consistently, and still not know what they were talking about or whether or not it was true? If the writer and you the reader may be forgiven for our ignorance. Even with this lack of clarity the Conference decided to accept quantum mechanics as a complete theory, recognizing that it gave no explanation of what the world was really like and that it predicted probabilities and not events. It was accepted because it correctly correlated experience, which the pragmatists held was the fundamental duty of science. The Copenhagen interpretation of quantum mechanics states that it didn't matter what the theory was all about. It was important that it work in all experimental situations – correlating what we observe under specific situations. The result was the statistics of probability.

Still today, the Hubble telescope explores the macro universe while the electron microscope ,the micro universe – with no way yet to look at what connects the two.

A year later Paul Dirac, an English physicist based on the early 19th Century invertible Fourier mathematical translation process, developed the Dirac Delta Function. It was a Quantum specific version of the translation that linked the wave and particle. It allowed a graphic representation of the collapse of a wave function in its transition from possible to actual, from the wavés super position state, in which all possibilities exist, to the spiking of a discrete particle. The following graphic modelling demonstrates that either amplitude or location can be measured and at the same time,the combining of the two states demonstrates uncertainty.

The Oscilloscope model of a wave function collapse

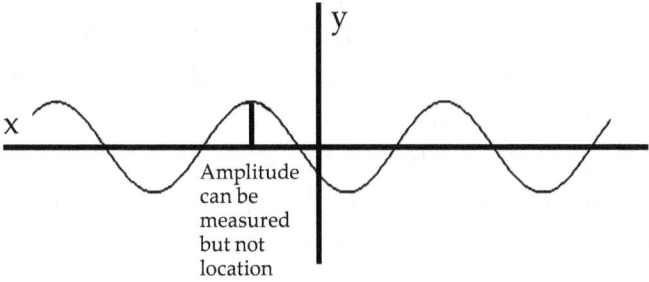

Amplitude can be measured but not location

Schrödinger wave. Super position state where all possibilities exist with no focus on any one – a chimera

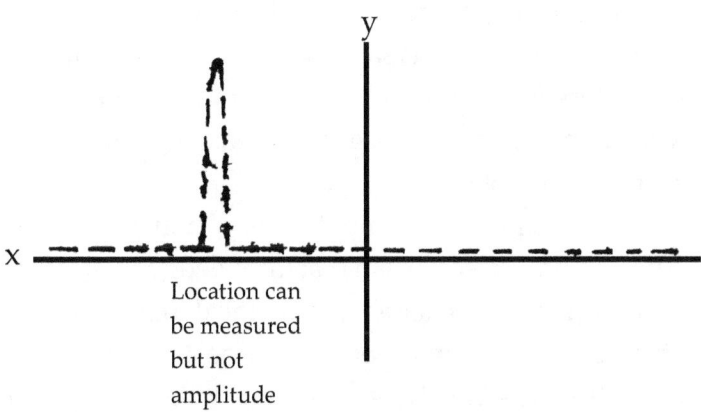

Location can be measured but not amplitude

Heisenberg Particle. Discreteparticle spike responding to the either/or real world situation

Quantum Theory

In 1929, Heisenberg and Max Born, looking for a resolution of the dilemma, developed a new mathematics that permitted description of Quantum phenomena in terms of either waves or particles. Mathematics is a language that allows physicists to talk to each other, if to no-one else. It is used in Quantum physics because its equations define more accurately than any mechanical means the fundamental phenomena beyond the range of vision. They work, even though not really understood.

That same year, Einstein in his pursuit of the scientific explanation for his absolute universe, published a Grand Unified Field Theory (GUT) but soon rejected and withdrew it. Twenty years later he proposed a revised theory that unified the two then known forces, gravitation and electromagnetic, proposing a set of universal laws designed to encompass interstellar space and the field inside the atom. Even with his ambitious goals it wasn't satisfactory. Science still looks

In 1935, an experiment was published that, although its effect wasn't acknowledged, spelled the death of the exclusivity of local realities and introduced the "emergence of non-local influences". This was the Einstein-Podolsky-Rosen (EPR) thought experiment which once again challenged accepted "truth". The experiment demonstrated an unexplainable connectedness between particles in different places. One seemed to know instantaneously the activity of the other. In the experiment two particles are sent off in different directions. One is sent through a magnetic field and experiences a particular spin, say up and down. The other instantly experienced the opposite spin. Then if the experiment observer changed the orientation of the magnetic field by 90 degrees, the first particle experienced a new spin,right or left, -the other, the opposite.

The EPR experimenters wanted to convey that in spite of the Copenhagen Convention, Quantum was not a complete theory. It did not include the possibility of certain physically real, but not observed, aspects of reality. They didn't seem able to explain how they knew this. Anyway, their argument insisted that for this apparent connection, all spin possibilities would have to exist simultaneously in the second to permit each of these spins to happen. Not only that, Quantum stood on the ground of probability, not causality. Since the theory could not speak to this situation, they deemed it incomplete. Bohr and Heisenberg disagreed.

Contrary to accepted physics, the experiment implied that information could be communicated at super luminal speeds, i.e., Faster than the speed of light. Einstein insisted it was not possible for an activity here to affect action somewhere else. He argued for local cause as common sense. The idea of an equivalent to telepathic connection was to him entirely unacceptable and for the time being held the day.

His colleagues had a different interpretation. For them the experiment particles were connected in a way that transcended the usual idea of causality, i.e., local cause, which held that what happens in area B does not depend on what an observer does in another and separated area A. The EPR experiment's implications was that either Quantum was wrong or that in fact, there was that unexplained super-luminary connection. Science theories of the time had assumed a basis of local influence and a maintenance of a state of separation between the observer and the experiment. There was to be no connectedness. No matter what was made of it, the EPR experiment could not be denied. There were forces of influence that could not be explained in terms of then current positions or belief. It was a classic example of the Gödel Incompleteness Theory which states that one cannot use the tenets of a system to prove itself -one has to go outside the system.

I have dealt in such detail with the EPR process because its findings predicated a major shift in the classical belief in separation. In 1964, J. S. Bell published a complex mathematical proof that, based on Quantum theory,(which presents statistical predictions that are always correct) showed that our common sense ideas about the world are inadequate and implied that at a deep and fundamental level, the separate parts of the universe, microscopic and macroscopic are connected in an intimate and immediate way. It demolished EPR's holding to local causes and enshrined the death of local realities and the emergence of non-local influences. As I hope we will see next in the development of Chaos Theory, ideas of separation were no longer appropriate. Everything was connected but not in ways we have been used to.

> P. S.: Quantum entanglement also speaks to "what if the parts were never separate?"

Chaos Theory

What we have assumed to be truth is subsumed in our belief in causality and linearity. e.g., the sort of skinnerian relationship seen when button "A "is pushed and "B"happens. As we begin to consider chaos this kind of predictability is no longer possible. For me, to get to where I am currently with the subject has required a major disorientation – sort of how the Newtonian crowd must have felt when faced with relativity (see the chart on Pg 41).Over the first two or three decades of chaos development, mathematicians and physicists, even if they had heard of this new theory, which was doubtful, paid little attention to it. You see, for them it wasn't physics and the math it used was weird.

    The first expectation I had to let go of was that I was going to discover a simple linear expression of what chaos is. Because it presumed an ability to describe the subject in terms of "is,"that turned out to be an inappropriate expectation. As I hope you shall see, non linearity is not that simple. I plan to look at some of the ways in which chaos expresses itself and intersects with other realities, not with the intention of finding any singularity but with the hope that from these various metaphors, the reader might infer some sense of wholeness. I hope to provide a slightly unfocused overview – like unfocused seeing, not looking directly at the object – like the mixed reality of a picture viewed very close, seen only as dots moving back for clear view and then way back where the image is without detail.

Complexity theory, nonlinear dynamics, has demonstrated that there may not be any chaos (true disorder) in the universe. Chaos means hidden order – you can't see it. That which appears to be incoherent, actually has an inner coherence with inner organizing patterns. The apparency of disorder is a function of the limits of our powers of perception and Fisher Information.

Complex Dynamical System Theory, Complexity Theory, owes its evolution to Ludwig von Bertelanffy, the grandfather of Systems Theory. In contrast with Norbert Weiner's "Cybernetics "which, as in regular tennis, prescribed net height, boundaries and rules, Bertelanffy said "just start hitting the ball i.e., initial conditions) and see what happens"No rules. He contended that everything was connected and through interaction of a system, the laws that govern the system emerge (as a function of initial conditions).

Years later Complexity is described as "the study of deterministic events in space and time that are possessed of the twin properties of emergence and irreducibility and are inordinately sensitive to initial conditions.

Chaos Theory 71

- Emergence is represented by islands of stability, i.e., appearance of order out of chaos (emergent chaos). One might say that emergence violates the 2nd law of Thermodynamics, – i.e., in a closed system things move towards disorder. However with Hubble's "red shift"finding, we know the universe is not a closed system.. In fact, a universe defined by Quantum Entanglement, in which everything is connected to everything, is suggested – a replacement for Heisenberg Uncertainty. Once joined, never apart.

- Irreducibility brings with it non invert ability, in which qualitative differences are not transferable. For example, a log cabin is not a small mansion, a mansion cannot be reduced to a cabin and a government cannot be turned into a business – also vice verse.
  Indent and built sensitivity to initial conditions is a key factor in these systems. Outcome is absolutely dependant on these conditions (e.g. A boat heading for Hawaii but off course one degree, can never get to its goal.)

Activities in nonlinear systems are deterministic over both the long and short term but only predictable over the short. It is through the sensitivity to initial conditions that we are able to say "events are deterministic". Changing initial conditions, whether regarding a physical system or a social system's mythology will change everything but will not predict where or when. Regarding global systems it is virtually impossible to know the initial conditions and their interaction with other global conditions.

The fundamental value of Chaos theory is seen in how it allows the prediction of when unpredictability will occur – but not what will occur. This information is valuable because it encourages the use of the precautionary principle, realizing that "Beyond this point there be dragons". For example, some California Chaolo gists knew that sticking a thumb into what was then a stable middle east would produce great instability.

It is through the use of computers that all the explorations into chaos have been possible. Before telling those stories, I want to introduce some of the definitions, ideas, and structures that have evolved – a cast of characters.

First, three bodies of study that are presently subsumed in Complexity Theory:

1. Catastrophe theory – the study of nonlinear thresholds and the straw that breaks the camel"s back. Looking at nonlinear phase change, Rene Thom gave insights into how apparently stable systems can suddenly transform. He found a way to represent, as a whole, non-linear dynamical systems,whether chaotic or stable and so complex as to be unpredictable.

2. Chaos Theory (originally derivative) allowing for the disappearance of perceived order.
3. Emergent Chaos Theory (integral chaos) allowing for the appearance of perceived order.

Of these three bodies, four main chaos models have been located, resulting from iteration of initial conditions:

1. Transient – a beautiful attractor (i.e., graphic representations in phase space) develops after missions of iterations and then suddenly drifts off, never to be seen again.
2. Transitive – a system, like climate, expresses itself with periodic behaviour (i.e. repeating patterns) and then becomes chaotic (e.g. global warming) and then returns to its initial pattern.
3. Intransitive – a system is zipping along and then suddenly goes into disarray because of some external influence. Assuming a cause outside the system, it is the choice of the "global warming "exponents.
4. Almost Intransitive – similar to above but nothing is done to cause it. Disarray is just the result of iteration. e.g., an irregular drifting in and out of long Ice Ages. With no cause, science can't use it.

Each of these models are subject to the influence of attractors. Attractors are graphic metaphors, all of which exist in phase space, an imaginary mathematical space based on √-1 in the complex plane. It is a multidimensional space, useful for graphing nonlinear equations, thus providing a means of turning numbers into pictures. In phase space, the complete state of knowledge about a dynamical system at a single instant in time collapses to a point. That point is the system at that instant. Pictures can be created to give an approximate mapping of all the instants in the system's history. Any attractor is a graphical representation, in phases space, of all the possible states of a dynamical system and which attracts possibilities of what can and can't happen. It is worth while to mention here that, in the superposition state, Quantum also presents all possibilities in the arena of the sub-atomic. Attractors are operators. They never repeat exactly because initial conditions never repeat exactly – however, similarities will allow recognition. As an example of an operation of an attractor, consider the functioning of Waddington's Chreode:

> Being a gully, a chreaode is an attractor. Through gravity, rainwater will always flow to the lowest point. During that flow the shape and surface of the gully will change (Hence, you can't step in the same gully twice).

So far three types of attractors have been identified, each representing its own version of recursivity:

1. Point Attractor – equilibrium (helix) – for example, a simple pendulum, the movement of which is always drawn to the energy death point, the singularity of silence, as it steadily loses energy to friction.

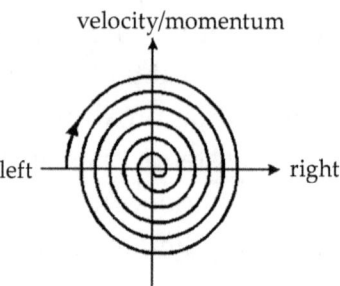

Phase space mapping of standard pendulum

2. Periodic Attractor – cyclical (ellipse) for example, a forced pendulum that receives a periodic kick at some related frequency, say every 2nd cycle (a period doubling bifurcation), after which the system loops twice before exactly repeating itself, creating a new rhythm at half the original frequency.

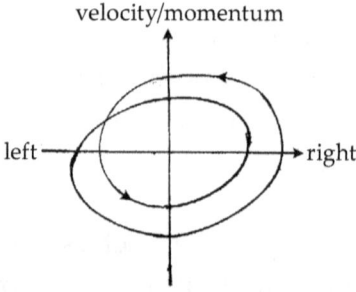

Phase space mapping of forced pendulum

Note: Both Point and Periodic Attractors are graphical representations of all the information about each particular system. Iteration creates that information in phase space in the form of Julia sets, i.e. completely bound, mathematically and geometrically. These attractors are neither chaotic nor self-similar, unlike strange attractors.

3. Strange Attractors – chaotic (fractal) as soon as you say chaos, you are talking about strange attractors, all of which are fractal. They begin with iteration of simple initial conditions.

a) Exactly self-similar – Mandelbrot attractor – a mathematical and geometric metaphor that is infinitely recursive.

b) Regionally self-similar-Lorenz and Hénon attractors (sometimes statistical) – never repeating itself but is infinitely recognizable.

c) Statistically self-similar – Rössler attractors (sometimes regional) recognized through its mathematical patterns.

N.B. Only a small number of attractors are fractal and all are embedded in the Universe of attractors.

The similarity between a man and his metaphor could be classified as regional in that one approximates the other. In a comparable way, current star patterns are our temporary metaphor for conditions that are constantly changing – except for

Chaos Theory

astrologists, the results of emergence are ignored. Our diagrams of constellations are super imposed patterns of order selected from essential disorder- not a true statement of nature.

Chaos – a Different Approach to Measure

Even though I may be hazy about their full significance relative to the Science of Chaos, I want to look at a group of ideas related to that subject, how it manifests visually, how we see it, and how we might count it.

To introduce the spirit of this discussion lets look at a famous paradox. Zeno of Elea, a 5th C BCE Greek suggested that greater than the infinite progression of whole numbers is the possible infinity found in going from zero to One, half the remaining distance each time, but never reaching One. "You can't get there from here."

As Göedel would have said, any concept that is to be viewed objectively must be embedded in a larger metric. In the example above, the infinite number of euclidean parts is embedded in the space between zero and One, an infinite number of dimensions. In the world of chaos, fractals are attractors embedded in dimensions which we call fractal dimensions. There are an infinite to the infinite power number of fractal dimensions between 1D and 2D and between 2D and 3D, all embedded in whole number dimensions and euclidean geometry. As further examples, a flat piece of paper is seen as a two dimensional object into which can be embedded a one dimensional line, into which can be embedded a Zero dimensional point. Incidentally, the euclidean definition of a point is a sphere of Zero dimension. A fractal point also has no dimension but inside may have infinite volume, energy and mass all contained within zero surface area. (Remember this is all nonlinear). As a result of the zero dimension of a point, there are an infinite number of points in a one dimensional line, no matter its length, an infinite number of lines in a two dimensional plane, an infinite number of planes in a three dimensional solid, and an infinite number of three dimensional solids in a four dimensional Teserac – a hyper cube with a 4D hyper-surface and its interior, which are not visualizable, (and definitely beyond me).

In these examples we have been looking at the embedding of a system, or an idea, into a more inclusive dimension. A more comprehensive system. For example: 1D to 2D to 3D. It is interesting to contrast this process with Plato's metaphor for life. He referred to the 2D shadows on the cave walls, cast from the then 3D universe onto a less inclusive dimension. ie 3D to 2D. Later with relativity, this became a 4D universe, but still 2D shadows.

And again, in a parallel manner, we can now see ourselves like 3D shadows projected from a 4D physical universe in a (incl. Time) 5D universe.

The mathematician, Georg Cantor, 1845-1918, in trying to answer Zeno's Paradox, placed the question within the then new concept of systems being embedded in something larger- a potential progression beyond infinity, as we think of it. In doing so he predicted fractal theory. Using the Hebrew letter "aleph"he concocted a progression that embedded infinities in new greater infinities opening worlds beyond infinity.

Aleph null – $\aleph_0$ = infinity to the infinite power
Aleph one – $\aleph_1$ = aleph null to the aleph null power
Aleph two – $\aleph_2$ = aleph one to the aleph one power
 ad infinitum

In this transfinite world there is no end to this progression. One's "there'" is beyond the ability to conceive of an end.

A further example of infinities is seen in what is known as the Set of Measure Zero. As it is applied to Lorenz's butterfly wing attractor, an image that results from taking a Poincare conic cross section of the attractors' infinitely dimensional torus and representing it as being two dimensional. Each of the loops in the image results from an iteration of the Lorenz equations. A definable arc segment of one loop represents the whole torus and is a set of Measure Zero – i.e. there is no chance of finding on the arc the point that is the initial condition. We recall that a point has no dimension. Therefore, like Zeno our arc segment would consist of an infinite number of points, more empty space than points.

Chaos – a Different Approach to Measure

To demonstrate this version of infinity, suppose we assume a computer capable of an infinite number of calculations per second, the age of the universe as 10 to the 40th seconds (15 billion years) and a sequence of an infinite number of available universes, there would not be enough time to investigate all the points on our arc segment. The chance of finding a particular point on the arc is virtually zero and not in the statistically relevant range. Any multiple of points will measure zero.

Up to now, our examples have dealt with euclidean dimensions. In the mind's eye a fractal is a way of seeing infinity, beyond our linear expectations. A fascinating example is seen in the Koch curve, conceived in 1904 by a Swedish mathematician and generating an infinitely long line surrounding a finite area. Starting with an equilateral triangle with sides of length one, add to each side a new triangle one third the size which increases the original boundary by four thirds. To each of the resulting twelve sides add new triangles one third the new side size, once again growing the previous boundary by four thirds. Again and again on and on to infinity. (3x4/3x 4/3 x4/3....infinity) and yet the circumscribed area is finite, only slightly larger than the original triangle. Seen as a fractal phenomenon, the Koch curve has a fractional fractal dimension of not 1D, not 2D but of 1.2618.

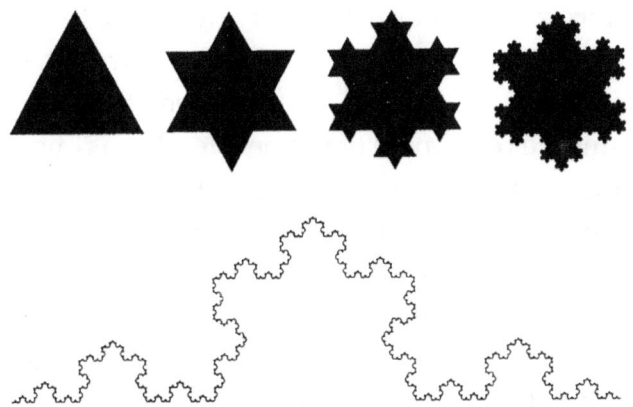

The Koch Snowflake

Above all, fractal means self similar, symmetry across scale, recursion and pattern inside pattern. The Koch curve, because of the required technique of constructing it at finer and finer levels, displays those fractal characteristics in a very visually understandable manner, even though hard to believe. As the line came close to infinity the quality of roughness – a fractal quality between 1D and 2D, but not dimension-able by Euclid.

If we add irregularity to roughness we begin to describe the fractal nature of a shoreline. Benoit Mandelbrot began to explore that subject when he asked the question "How long is the coast line of Britain?" You guessed it – virtually infinite, like Koch's curve. It became becomes a function of the chosen unit of measurement. Starting with a yard stick, it will obviously skip over the twists and turns, smaller than a yard. Next a foot ruler which will incorporate more of the irregularities, then an inch, then a millimetre so we can measure around pebbles or grains of sand. Then we need to take into account the roughness of the pebbles. Each time we reduce the unit of measurement we get a greater coastline length until finally the unit is an atom. Is this when the process comes to an end? So -how long is the coast line of Britain? Is it possible to see fractal y ? Is that what Escher was doing? This systematic reduction of the unit of measurement is also used in Time Dilation Theory.

So far Chaos has made no reference to truth. If any suggestion of it is to be see, it would be expressed in visual terms, relative, not absolute and understood through perception. Where Newton has pinpointed the particular discrete frequencies of each of the colours, Goethe contended that the reality of colour was in our perception of it. There was no real-world quality of redness, separate from our perception. Redness is a territory of a chaotic universe and except for the colour blind, we can all agree on it. The main point to be made is that, where the quantitative form of description has predominated in earlier science, with chaos it is about the qualitative.

The knowledge we can gain of a subject visually is relative to our perspective, our proximity to it. Consider this sequence of topological relationships:

A baseball seen from 1000 yds. Is only a point-IE, 0D. Getting closer one sees a circle i.e., 0D. Getting closer one sees a circle i.e., 2D. Closer again, one sees the shadows – ,i.e.,3D and right up to the eyeball, we are back to 2D.

Loosely only, fractal dimensions are topologically similar to this model, even though fractal dimensions are partial dimensions. Note also that going from 0D to 2D without 1, displays non-linearity.

Now imagine an ant on the ball. In each of the previous situations the ant would only see 2D – its just a matter of perspective.

A classic example of the limits of knowledge resulting from topological relationships is evident when we consider the 1970 moon shot of the earth. The image produced suggested apparent global stability, giving no indication of the unstable elements of Vietnam, floods forest fires, etc. These elements were like "butterfly wings"– conditions that were invisible from the moon but were influencing everything – even if we didn't know when. So in a chaotic universe, how and where do we get information – and should we expect it to be stable?

Some Early Chaologists

With his Butterfly Wings effect Edward Lorenz has been called the discoverer of chaos- its father. If that description was appropriate his role must be shared with a series of step-fathers. Chaos is primarily a visual metaphor. In that respect its conceptual parent is E.C. Escher.

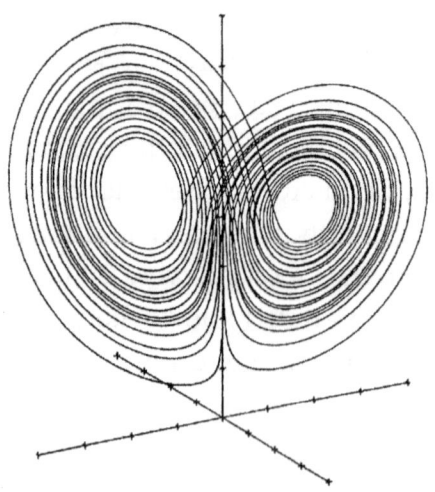

The Lorenz Attractor

His earlier works, particularly the stair, drawings that dealt with recursivity, presaged a new reality. Other of these parents included James Yorke, who inadvertently gave Chaos its name, Benoit Mandelbrot's computer iteration of simple mathematics produced and named the fractal, Mitchell Feigenbaum and his non linear constant ratio of intervals between sequential bifurcations and Dr Otto Rossler whose attractor, almost as famous as the Butterfly Wing, describes aspects from vulcanism on earth, Jupiter's moon IO, rabbit population birthrates to star cluster formation.

As a boy, Lorenz was intrigued by the subject of weather, its changeability, its patterns that come and go, always seeming to obey mathematical rules, but never repeating. At MIT in 1960, he began creating on his computer a series of toy weathers. He and his colleagues were fascinated with the way his created atmospheric patterns changed over time. Initial input consisted of twelve "laws of nature"numerical rules that expressed relationships between temperature, pressure, and wind speed in a Newtonian like deterministic sort of way. His twelve equations calculated over an over, mechanically –played with the patterns of weather. He loved the orderliness, the familiar cycles appearing repeatedly but never twice the same. One afternoon in 1961, something happened that opened a door to a new science. Wishing to check again one sequence of a previous run, he restarted it halfway and let it run. When he returned, expecting to find a duplication of the original, he found that the old and new patterns had diverged so rapidly that within a few months they were unrecognizable as being the same system. What was wrong? He soon realized that the figure he had provided as the initial condition, the starting point was 0.506 when in the original run it was 0.506127. He had assumed that one part in ten thousand would have little effect and he could save printout space. Well,m he was wrong. With his particular equations, small errors proved to be catastrophic.

Lorenz could have assumed that something was wrong with his model, his technology or determined his equations were a joke relative to the earth's real weather. However, given his inner affinity with the subject he had faith in his findings. That very day he decided that long range weather forecasting was doomed. He realized that weather must be inordinately sensitive to initial conditions -maybe a butterfly flapping its wings in Indonesia could effect the weather in say Brazil. He also realized that any system that acted non-periodically was unpredictable – forecasts beyond 6-7 days worthless.

Lorenz saw more than randomness in his weather models. He saw geometrical order posing as randomness. He turned his attention to the mathematics of systems that never reached a steady state – almost but never. How could such richness and unpredictability, such chaos arise out of such a deterministic system? He felt there had to be a link between a periodicity and unpredictability in weather - no repeating and no forecasting.

In a 1963 issue of the "Journal or Atmospheric Science"m, Lorenz published "Deterministic Non Periodic Flow". It is not surprising that no physicists or mathematicians concerned with dynamical systems would think to look in a journal of meteorology. It wasn't until 1972 that a fluid dynamicist came across a copy of the paper, loved it and gave copies to his associates. One of these was James Yorke, a young mathematician who liked to think of himself as a philosopher. It was what he had been looking for. Not just mathematics, it provided a vivid physical model, a picture of fluid, in this case air, in motion. He knew the physics community must see it also.

Actually, the year prior to Yorke's discovery saw Lorenz's first line mapping of his attractor, as an attachment to his, paper had been included in the scientific literature. To plot these first seven loops, the precursors of his famous Butterfly, the first strange attractor had required five hundred calculations on his computer. By 1963 he was using his three convection equations, the three variables produced the attractor in 3D phase space. Each loop represented an iteration of the equations, but never intersected each other. The attractor was strange – not periodic and only seven loops.

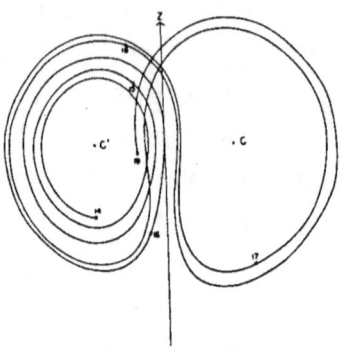

The First Butterfly

When in the early 70's, May and Feigenbaum had been working with the repeated iteration of the quadratic difference equation, no one knew that in 1964, Lorenz had been looking at the same thing. His results had suggested that the earth's climate might never settle reliably into an equilibrium with average long term behaviour (the goal of run of the mill climatology). His exploration was a metaphor for an unasked question "does a climate really exist?" Could it be long term averaged meaningfully? Was there any pattern to its behaviour? Why was the average weather for the past 12,000 years so different from the average for the previous 12,000. Were they just fluctuations in a larger pattern? In considering global warming, was it the Intransitive or Almost Intransitive model of chaos? We recall that the latter model would have climate drifting into a different form of behaviour without any particular reason or cause. Might that explain the Ice Ages as a by product of Chaos?

Climate is global, it always has been global. Do our attempts to describe it as some orderly pattern embedded arbitrarily in the dimension of time, do justice to the Science of Chaos? Surely emergence depends on initial conditions -and starting when? The classic question for Chaos.

Meanwhile other pioneers in this new science, each removed from the others, were making unexpected discoveries, in spite of the resistance of orthodoxy. Belousov and Zhabotinsky (B/Z) were working for the Soviet Academy of Science in the 50's. They found that a specific mix of particular chemicals always resulted in the same observable physical pattern. The Academy said it couldn't happen. B/Z replicated the experiment a hundred times. Even though they witnessed this replication the Academy said it violated the Second Law of Thermodynamics and discounted all the evidence completely. Even though chaos was accepted as a science in the mid 70's the B/Z results were not accepted until the late 70's. Being a pioneer is a real uphill voyage.

David Ruelle and Floris Takens, two European mathematicians, in 1971 published "On the Nature of Turbulence". Rooted in the mechanics of Point and Periodic Attractors, they had noticed, when looking at turbulence in fluids, that if periodic motion became unstable, there appared a geometrically strange object. They called it a strange attractor existing in phase space. It was not periodic and never repeated itself. By using mathematical reasoning, they predicted the characteristics of the Lorenz attractor. Ruelle, with no experience in fluid flows was not discouraged, stating "Always nonspecialists find the new things". Gödel would have been proud of him.

In many ways, these explorers didn't know what they were chasing – it was so fascinating, the just couldn't let go.

Another parallel investigation was getting under way in 1960 France. Micheal Hénon, an astronomer, was looking at the subject of globular clusters, dense with up to a million stars. Every so often, a single star will gain enough energy from an interaction with a binary to fly off and the rest of the cluster would contract. He questioned how these extremely complex clusters stayed together. The answer would have ramifications in understanding our solar system which appeared stable over the short term – but who knows?

In 1962, with access to computers while at Princeton, Hénon began modelling the orbits of stars around their galactic centre. The orbits were periodic but not completely regular, never repeating exactly. To plot the computer patterns formed by the orbits, he imagined a flat sheet placed upright at one side of the galaxy through which each orbit would pass and be plotted. The resulting mapping was a cross section of the 3D torus, or doughnut, that represented the stellar galaxy. The initial plot produced an egg-shaped curve, which, as the input increased-became more complex loops, crossings, unstable orbits with randomly scattered points, not fitting a curve. The picture became quite dramatic, indicating complete disorder mixed with the clear remnants of order. Hénon could only explore and speculate. In 1976 he heard a talk about the Lorenz attractor. The speaker, using the appropriate differential equations, had been trying to illuminate the fine detail of that attractor but with little success. Hénon had an idea. Rather than differential, he would use difference equations, discrete in time, that would allow repeated stretching and folding of phase space, like a chef making croissant pastry. Starting randomly and using coordinates derived from Ynew=0.3x and Xnew=Y+1-1.4x2, he plotted five million points. At first they jumped randomly about the screen, but quickly what came to be known as his banana shaped attractor began to emerge. The more the program ran, the more detail was seen.

Initial Hénon attractor

What seemed to be single lines, on magnification became pairs of lines, then pairs of pairs etc, displaying infinite recursion. As an attractor it is the trajectory towards which all trajectories flow and converge, without being dependant on any particular choice of initial condition. Even though poorly understood by mathematicians, anyone can achieve the results on a personal computer.

The biologist Robert May, in 1971, became involved with his Princeton peers in his special interest – a study of how single populations behave over time. He knew the classical linear Malthusiann equation (Xnestrx where x is population [0-1] and r is the rate of growth parameter) had limited value. It made no allowance for conditions or restraints such as food supply or competition. What happens he wondered, when that nonlinear rate of growth parameter passed a critical point. A modification of that equation, known as a logistic difference equation (Xnest = rx [1-x]) kept the growth within bounds since as x rises [1-x] falls. The rate of growth parameter would represent all varieties of messy non linear characteristics like population fecundity, competition, etc. How did these parameters affect the ultimate density of a changing population?

Using the latter simple equation and an assumed fish population, May investigated hundreds of different values of the parameter. When low, his model settled into a steady state. However, when high, the very same system seemed to behave unpredictable. What happened at the boundary between the different kinds of behaviour? Suddenly, as the parameter passed, the line of his mapping broke in two. Like a Y it bifurcated. The imaginary fish population oscillated between two values in alternating years. Turning up the parameter a bit more produced a string of numbers, each returning every fourth year. First the period had doubled from one to two years, and then again from two to four. Beyond a certain point periodicity gives way to chaos, oscillations never settled down.

In 1972, James Yorke, a University of Maryland mathematician, was given a copy of Lorenz's 1963 paper. "Deterministic Non periodic Flow" He was beginning to realize that in spite of the search by physics and mathematics to discover regularity, disorder was real and needed to be understood if it was to be dealt with. He was familiar with the work of his friend May and the system behaviour he had exposed. He analyzed this behaviour with mathematical rigour in his published paper "Period Three Implies Chaos. "It verified Mays work and showed that using a logistic equation, it was impossible to set up a system that would repeat itself in a period – three oscillation without producing chaos.

Mitchell Feigenbaum began to think about non linearity while at Los Alamos. He knew of May's work and in 1975 heard Steve Smale talk about the mathematical qualities of quadratic difference equations and about the point at which mapping changes from periodic to chaotic.

Feigenbaum felt he could look further. He knew that on the route to chaotic behaviour there was a series of period doubling s – one into two, two into four, four into eight, etc. He used an ananalogue of the equation May had used in his population studies, a quadratic difference equation ($y=r[x-x2]$ where x was 0-1.0. He knew that the bifurcations, the splitting of the mapping lines, took place at points where there was a slight change in, for example, the fecundity parameter value in a population of gypsy moths. He calculated the exact parameter values that existed at each period doubling. As he wrote down the figures from each iteration of the equation as read from his hand held computer, he began to notice a regularity. The numbers were converging geometrically as each period doubling took place something in the equation was scaling. He was able to calculate that the ratio of convergence of the value of the first to the second, then the second to the third was consistent. The ratio turned out to be 4.669 and seemed to have no relationship to the standard constants -it had a life of its own. Robert May had also noticed this convergence in the early 70's, but having a different metaphorical lens and not being a mathematician took, little notice.

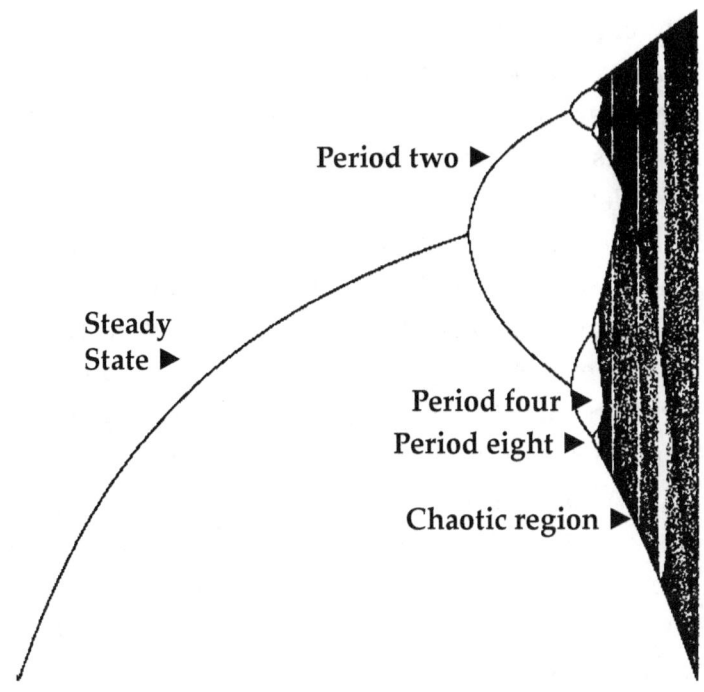

Period – Doublings and Chaos (Bifurcations)

Feigenbaum then computed the doublings for a sine equation [$x_{t+1} = r \, \text{sinc} \, \pi \, x_t$] and found the same ratio, identical to the regularity of the previous simpler function. He tried other functions, each of which went through a series of bifurcations on the way to disorder. Each one also produced the same ratio. Something in these simple equations, when repeated over and over, produced the scaled sequence of the same number, 4.669. He had stumbled onto something he referred to as universality, but commented "the only things that can be universal, in a sense, are scaling things," proportion-

Some Early Chaologists 97

ally based. This meant that different dynamic systems would behave identically. He believed that his theory expressed a natural law, but non linear about systems at a point of transition between order and turbulence.

Benoit Mandelbrot, a European mathematician, was associated with a 50's Paris community of purist mathematical formalists. Purist is to say, it was numbers, not geometry. This was a problem for Mandelbrot. A new reality was forming in his mind, a reality to evolve into a full fledged geometry. What was important to him was the computer's visual display of that developing geometry. His peers thought he was crazy and he felt forced to flee his Paris community. He accepted the shelter of IBM in America, where he stayed for thirty years.

In 1960, working out of their pure research lab, Mandelbrot had been dabbling in economics, with particular interest in the change of commodity prices. Conventional thinking at the time held that small, short term transient price changes, which it was believed, were determined by deep macro economic forces. Mandelbrot had no interest in this dichotomy. He was looking for patterns that linked the two classes of influence. He was convinced that there had to be a symmetry across scales. In his study of cotton prices, though each change was random, the curves for both daily and monthly changes in price matched perfectly. In fact, he found that the degree of variation had remained constant over the sixty year period of study. He moved on, taking with him his belief in the significance of scaling.

In the mid 70's, though the quantumists and relativists were still fighting, others were noticing other things – patterns in nature, particularly in noise, leading to an idea that some order, like maybe music, might be hidden in the presumed disorder that noise represented. Mandelbrot found some thing else. Along with his IBM study of commodity prices, he began investigating the "noise" that was resulting in electronic communication error. He was trying to understand the nature of intermittent static. And he was aided by a construction (or destruction) that had evolved from Cantor's transfinite theory. It was called the Cantor set, an intermittent structure. Starting with a line of any length, remove the middle third, then remove the middle third of the 2 remaining lines. Continue removing thirds from the remaining, again and again, ad infinitum. What remains is a dis-continuum, a dust, an infinite number of points with a total length of zero. The Cantor dust actually had a fractal dimension between 0D and 1D, 0.6309.

In looking at his noise problem, Mandelbrot found a similar dis-continuum,. The static had layered within it periods of silence. At finer and finer scales the remaining noise had gaps of silence. Intermittency had a fractal structure across scales. At every time scale, hours to seconds, the relationship of noise to silence remained constant.

Some Early Chaologists

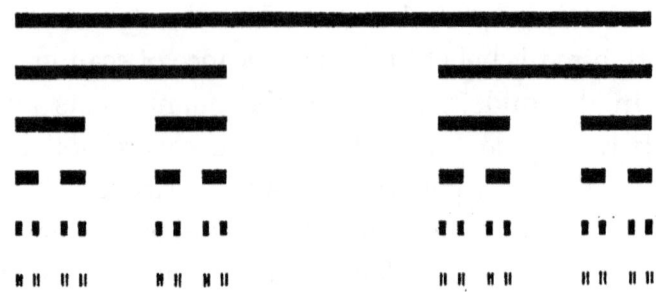
The Cantor Set on the way to Dust

Mandelbrot was greatly interested in the idea of dimension and how to describe irregular shapes, not describable with the legacy of Euclid. We have already discussed the concept of the fractal, named by Mandelbrot in 1975, and fractal dimensions. Fractal units not bound by euclidean length, width, and height permitted for instance, the more exact measurement of a coastline. In that exercise he held that the information transfer was a function of the relationship between the object and the observer. More than plain mathematics he claimed the influence of relativity. The position of the observer affected the measurement. However there was a weakness in this understanding. You will recall the exercise of sighting a baseball from different perspectives, the transition of reality from one to two to three and back to two dimensions. In actual fact there were no clear boundaries where those transitions took place. However, understanding what the perception of the "in between state"might be, helped open the door to his reality of fractal dimensions.

Situations like shoreline measurement and the Koch curve made little sense to formal mathematics. Mandelbrot, however, was a geometer and saw form as something drawing could describe. With access to the IBM computing resources, his intuitive imaginings could be drawn. As a result an important characteristic was becoming clearer. Above all, fractal meant self-similar, symmetrical across scales. The Koch curve was exactly so pattern - within pattern at finer and finer scales until one finds oneself measuring fractally the irregular surface of apparently completely smooth materials. He found fractal structure and bifurcation everywhere in nature. In the growth of trees,in our lungs' air passage network and in our blood circulation system. In the latter case, blood and vessels take up to 5% of the body's volume and yet keep all cells no more than three or four cell's distance from a vessel – what a system!

Recognizing such complexity in natural systems was one thing but creating it on a computer screen was another. It was believed by some that it was the iteration in the complex plane of particular equations that gave rise to strange attractors. John Hubbard and Micheal Barnsley explored and mapped, in the context of dynamic systems, various shapes that reflected the behaviour of forces in the real world.

But it was Mandelbrot who was on the trail of something special. He was to discover the ultimate fractal shape - the Mandelbrot Set.

Some precursors of this set began to appear when he was trying to generalize those mathematically bound shapes in phase space that had been invented in France sixty years earlier. They were referred to as Julia sets.

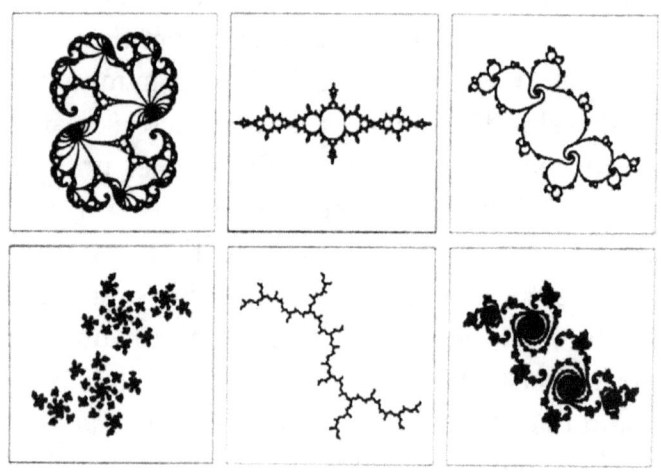

An Assortment of Julia Sets

In 1979 he created an image that referenced, like a catalogue, each and every Julia set. As he pressed the limitations of then current computer capacity, he had no idea what was lingering hazily behind the screen. As he refined his mapping process, the first outlines of discs appeared, and then hints of more attached shapes. What then began to look messy, sprouts and tendrils reminiscent of the Julia shapes turned out to be a real part of the development of his own set. At greater magnification and finer detail, smaller but exactly, self similar shapes would repeat. The successive smaller discs scaled with geometric regularity -it was Feigenbaum's ratio of bifurcation.

The Mandelbrot set is a collection of points, i.e., complex numbers, either in the set or not, that results from a simple iterated calculation, in the complex plane, of the mapping Z ---> Z^2+C. Take a complex number, square it and add C, a constant. The result becomes the new Z. Continue. Each dive, deeper and deeper into greater magnification, produced new surprises. This attractor has been described as and appeared to be, exactly self-similar across scales. However, with greater magnification, none of the tendril or sea horse shapes exactly match any other part, even though obviously related.

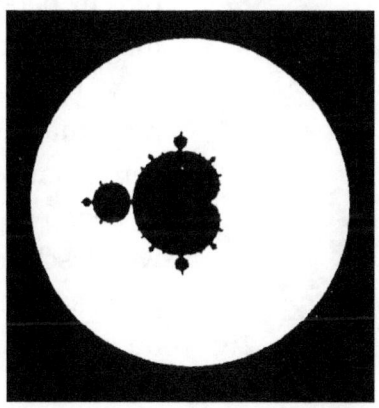

The Mandelbort Attractor First Appears.

The parallel claim for the quality of infinite recursion is very difficult to challenge. The publication "Turbulent Mirror" by Briggs and Peat includes a

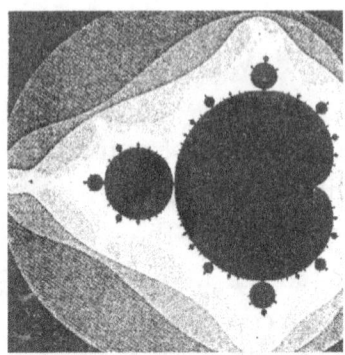
Natural Sized Reference

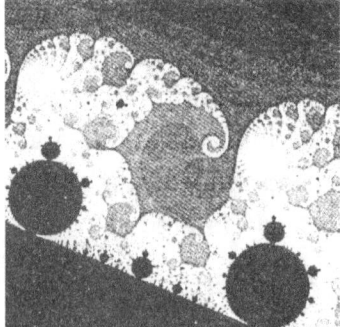
Order Floats in Chaos

2.7×10^9 *Magnification*

series of eleven images (3 shown here) tracing this quality as the Mandelbrot set goes deeper, displaying at each level of magnification its unimaginable riches. The top image is the initial clear appearance of the attractor in its atmosphere. The next represents a multi-million sized magnification and then 2.7 billion magnification. To put that in perspective, the magnification necessary to produce a 10 mm image of a hydrogen atom would need to be 0.6 billion. When we describe this process as infinite we have no reason to think this recursion would not go on forever. Its just the nature of fractal order.

One has to wonder where these fractal components, perceived as infinitely tiny, displaying correspondingly tiny detail actually come from – and forever. It is not just material. As Hubbard stated. "There is no randomness in the Mandelbrot set. It obeys an extraordinary degree of leaving nothing to chance".

Again I ask, before its emergence, where had it existed – on what plane? Perhaps this is the way of Nature, using simple physical laws to iterate itself and out of that process created an orderly universe – all this suggesting a non - linear process of evolutionary development.

Chaos Applied

The study of Chaos has primarily been an attempt to understand complexity and thereby make use of that approach. The Chaololgists, emerging from a world that believed simple systems behaved in simple ways and complex behaviour, implying complex cause, found an opposite situation. Simple systems behaved in complex ways and complex systems could behave simply. Most importantly, the laws of complexity hold universally, without regard for a systems atomic composition. A good part of the scientific community had heard of something Called chaos, but most just carried on in their original orientation. Those relatively few with a focus on Chaos could not agree on the details but each, with their own metaphor, knew very clearly what they were doing – studying complexity, and they were able to do it without the rancour seen in the Relativity/Quantum feud.

Traditional science with its traditional methodology looks to establish an ordered environment that controls conditions in which an hypothesis can be tested and the results replicated through experiment. In Chaos, order cannot be imposed. It must emerge out of the dynamic system. As we know, what emerges will be a function of the initial conditions, what we have described mythologically as beliefs. To influence emergence we just change the conditions. Chaos theory says that changing belief will change everything but not foretell where or when. We must find and acknowledge patterns that will indicate emergent qualities – certainly not the traditional application of the Scientific Method.

Depending on objectives, both order and disorder can be encouraged to emerge. Ilya Prigogine, when he talks about "dissipative structures," refers to systems that lose energy to friction and display instances of self-organization and disequilibrium. They take advantage of entropic disorder while chaotic order emerges. Dissipative structures address a minimum sustainable threshold argument that is applicable to either desirable or undesirable situations. Attractors act as dissipative structures. Even if the attractor applies to the whole system, dealing with only part of it can arrive at a minimum sustainable threshold, – i.e., a tipping point or phase change, such as in these following examples.

ORDER OUT OF DISORDER

With regard to world starvation, assuming that the most extreme condition in Africa allows $1 per person per day for food, even though still insufficient, raising that figure to $1.50 per person per day might reach a tipping point and change the whole situation.

DISORDER OUT OF ORDER

Prior to 1981, international intelligence understood that the USA was spending 10% and Russia 35% of GDP on arms – an assumed maximum sustainable threshold for Russia. In 1981 it was learned through the spy network that Russia was actually spending 65% – barely sustainable, beyond which the centre could not hold. Reagan decided to push things. The idea of star wars was initiated, moving the metaphorical battlefield into space. US expenditure was raised to 15%. Russia couldn't respond. The pressure plus Chernobyl and the Afghan war brought her to collapse.

Changing the Metaphor

No surprise - we humans unknowingly restrict our progress. Our organizations continually seek to reconfirm and validate their current existence. This effort leads to a selective use of information and a dangerous narrowing of perception. The potential evolution of new knowledge systems is thereby stifled in the attempt to perpetuate equilibrium. The process gets stuck.

The results of this tendency are expressed in the maxim, "That which begins as heresy, can often end as dogma." New ideas, even though more effective than currently employed are often seen as heretical by the establishment. If and when adopted, they, in spite of further new ideas, tend to be maintained in the interests of efficiency. This iteration, back and forth, effective devolving into efficient, is the ongoing human ritual, only to be modified by changes to the metaphor. Some examples follow to show how being stuck in a particular metaphor does stifle the possibility of moving from efficiency into the greater value of effectiveness.

Carl von Clausewitz, in his exploration of military strategy, claimed that that possibility was limited by three factors:

- Uncertainty - confusion as to what to do
- Internal friction – the result of doing things as they were done in the past – its efficient
- Inflexibility – the tendency to keep responses the same.

Exercising flexibility allows the emergence of a new metaphor that reduces internal friction and results in new certainty and effectiveness. That emergence produces a new way of seeing things and the accompanying move to greater effectiveness.

Remember the Maginot Line? Prior to 1940 the French built that supposedly impregnable barrier on their northeast border with Germany, on the assumption that the enemy was there and that was the border that needed protection. The German army didn't have the same metaphor. By taking a different route through Belgium, they simply went around the fortifications. Another disjunction was evident in the different use of tanks. The French army saw tanks as support for the infantry. With no similar belief, Germany's ' tanks though inferior to France's were not restricted to foot soldier speed. As a result, they were much more flexible and won the day. The Maginot line just gathered moss.

In the 1920's Billy Mitchell, a pioneer in the US Army Air Corps, realized that air power trumped battleships and that carriers were the means of delivering that power. The traditional battleship

Navy had no interest. Around 1925 he announced that Pearl Harbour would be struck on a Sunday morning by carrier based Japanese Planes. The Navy considered him crazy. He was given a chance to demonstrate what the Navy said was impossible. He sank, from the air, a captured German ship. The Navy still wouldn't believe it was possible. He was court marshaled for suggesting planes could sink battleships and for insulting their ally Japan. Also we recall the resistance Belousov and Zhabotinsky faced.

Another flyer, after returning from service in Vietnam, saw the possibility of a new market in the mail delivery business. When he suggested in his degree dissertation that all mail could go to a central hub, in this case Atlanta, before being sent to their various destination, his professor failed him as being impractical. It would be an unnecessary step, more expensive than what was provided by the Postal Service getting the mail to its goal, when it could. In this case it was efficiency in the form of being economical.

However our flyer and his buddies had a different metaphor. They knew that this new market would be quite willing to pay many times more, particularly when delivery was guaranteed overnight or the company would pay the cost. Thus emerged Fed Ex, soon to be copied by U.P. S. The professor is not remembered.

The last example again demonstrates the power

found in new metaphors if well chosen. In 1980, IBM hardware was huge, based on their belief that computers were only for "Big Business". Meanwhile, the Apple ∏ had arrived on the market. Designated as a home computer, it was affordable, unique but relying on Doss at the time was not very effective or efficient. Nevertheless it was seen as a technological art object. Sales took off. By 1982 the designers were millionaires. It took the IBM bureaucracy another year to recognize that they were missing the boat.

1984 saw a major phase change. It was the intransitive model of chaos. Apple introduced Paint Pro, a tremendous boon for artists and designers. It worked so well only because Xerox Park had sold them "windows" and "the mouse" for one dollar – for after all, Xerox made their money from copiers - right?

The next step saw the birth of MacIntosh and printer, both with colour and enough computing power to run Paint Pro. To cap this initiative they put their entire advertising budget into a stunning T.V. commercial to play multiple times during the Super Bowl. The next day, their stock hit the roof. More and more designers recognized their value. The private publication market opened.

By 1985, Apple had overtaken IBM, who finally got around to producing the "personal computer"which took over the Business and Game market. Microsoft continued to benefit. Bill Gates had made a deal whereby every IBM, clone or Apple computer sold had to pay $50 for the Windows operating system, even if another was used. The american dream.

Demitrios Dendrinos, an architect/scientist demonstrated the value of being able to see things differently when he wrote about "Cities as chaotic spatial attractors."In this metaphor he presents an alternative to the normal grid or other geometrical road pattern that north americans, with no local history, have traditionally imposed on their settlements. These patterns, though efficient, had no relationship to the existing topography or attractors. There is a corollary found in modern freeways, slicing efficiently, straight through a rolling landscape. We can compare that approach with an early simpler roadway, staying level as it effectively responded to the topography, gently and curving, without major cut and fill. The existing physical characteristics are the attractors that can determine the form, almost organically. Dendrinos was only concerned with point and period attractors, nothing strange. Although he determined his design forms in computer phase space, the results looked exactly as they should.

In each of these examples, success came from seeing things in a new way, with a new metaphor. As a result, new ritual seems natural and even obvious, as long as the process of discovery is allowed. Carl Jung said it well ---"Because the world itself is not so much as it is, as how we see it."It would seem that effectiveness is best achieved by assuming that each and every situation is unique to its moment, determined by its applicable initial conditions and selected metaphor.

Success is dependant on getting these factors right. When our understanding of a systems operation is based on incorrect assumptions of initial conditions and supposed linearity that understanding will not be useful. In such situations there could be real value in assessing the appropriateness of those assumed conditions. At least the assumptions could be challenged. If errors are discovered and corrections made,the door is open for a new metaphor, based on real world.

Recall that a metaphor is a translation of a myth or belief -the operating initial conditions. Whether on the scale of individuals, groups or cultures, we ourselves are the fundamental initial conditions generating the applicable metaphor in our individual or collective vision of life. Each of us is looking through our particular metaphorical lens. No one else exactly duplicates what each of us sees. However, like the blind men trying to describe an elephant; after each has touched separate parts, the total of our collective experience and metaphor does describe the whole. We generally don't realize that what we see is a function of what our own metaphorical lens tells us. If we did realize this, we could easily change our focus. This is particularly valuable if we learn to see with anther's eyes and then provide a new metaphor that accommodates both the new conditions and the others' belief.

In the case of the global warming debate that currently favours a linear based model, what if it was reframed around a different metaphor? Important questions would emerge. Climate is obviously global but does it exist in any orderly pattern embedded in the dimension of time that we attempt to impose on it? Could it in fact be a non - linear chaotic model, and if so, transitive, intransitive or almost intransitive? And which initial conditions would we be looking for -where would the starting point be? One can see how these questions challenge the scientific community and why it is generally not involved in that debate.

Recognizing Chaos

Looking back at the preceding examples, it is easy to see how the changes in metaphor may have often resulted from an intuitive hunch. However, when we look at the results of the change with hindsight, they often seem obvious. Just a matter of seeing things differently. Starting with the understanding that not every situation is chaotic and knowing the value in seeing where unpredictability could appear, it should be possible to scientifically identify where chaos exists – and it is, in fact possible to measure it.

First though, to recognize chaos, knowing it is both "strange and fractal", we can compare any particular system against a check list of characteristics. "Yes" indicates chaos:

- It is not predictable over the long term.
- Lyapunov exponents are positive, meaning information is being lost as order reduces.
- An inordinate and disproportionate estimation and dependence on initial conditions.
- Non linearity
- If logistic equations are involved, chaos is indicated. NB all G elements must be present. i.e.,

non - linear, first order (not quadratic), difference, limiting condition, topologically transitive (no exact repetition) and initial condition sensitivity.

Lastly we need to determine which scientific model applies: Linear strong, Linear weak, (both are Newtonian) Chaotic weak/soft (small outcomes and barely non - linear) and Chaotic strong(non - linear chaos as we know it).If our system turns out to be chaotic, we can move on to measure, in fact to apply the Scientific Method. You will recall when we compared the four main fields of science, we learned that, unlike the other 3, chaos had a qualitative, non - linear mythology that could only be expressed visually in phase space on a computer screen. Lets see what we find when we relate the Scientific Method to a typical experimental investigation, learning more about what seems to be a chaotic system.

For our example, we will assume a simple logistic equation is entered. As it is repeatedly iterated we may observe a particular graphing of points beginning to appear on the screen. The graphing process might resolve into a torus, for example, a doughnut shape. If a Poincaré cross-section (a mathematical graph) is taken of the torus, we may discover an attractor. Though still in phase space, this is equivalent to the rational analysis step in the scientific method. To describe the system more exactly, a resulting fractal structure could be qualitatively measured. Also, the Russian chaologist Kolomograov, developed a method of determining what he described as K-entropy in a system. It is a measure of disorder in the graphing of the Lyapunov exponent's value, describing the gain or loss of information and order within the system. This factor can be established. An hypothesis may be established.

After completing the run and its graphing, the experiment could be reproduced, the equation could be run again but with a slightly different starting point so that the degree of sensitivity to initial conditions can be determined. The previous measurements would be taken again. The Lyapunov exponent can be compared numerically and the fractal structure visually. We might recognize similar patterns, but e3xcept for Mandelbrot, never exactly and sometimes regional or statistical.

Once this phase space analysis has been verified the findings may6 be translated back into real world terms for application. An hypothesis could tell us when unpredictability could appear – or,by finding another planet "earth-like"in its similarity to our wondrous "blue Planet", we might be impelled to assume it could have a similar atmosphere, etc. Could that be useful information? Could our senses be trusted?

The case rests. Chaos is definitely a science. Although obviously non - linear, and using a strange mathematics and geometry, the steps of the Scientific Method are satisfied: Observation and Measurement, Rational Analysis and Hypothesis Formation and Qualitative Replication. Chaos is a very interesting individual corner in the room called Science, and in the mansion called philosophy, that penultimate expression of all human knowledge and wisdom.

There is a Buddhist saying that translates as Individual, Group, No separation.

We started this exploration at the time of Pythagorus. Regarding this quality of wholeness, what might he have held?

"the concept of a harmonious universe ordered according to the "great chain of Being"-the continuum of matter, body, mind, soul and spirit - stands as one of the fundamental ideas of Western thought. It is represented in countless examples of cosmology, literature, architecture, theology and art, penetrating deeply into our scientific and educational institutions and profoundly influencing thinkers from Aquinas to Kepler, and Newton to Einstein. All of this can be traced back to the enquiries and teachings of Pythagors"

"Divine Harmony" Stromeier and Westbrook.

Conclusion

When we look at the role science plays in the process of determining cause and making appropriate choices, we discover a distinctive characteristic offered by Chaos. Answering the question "why"or "what causes that?" is a pretty fundamental process for traditional science. The answers are usually based on history, what happened before, and having that information we are linearly able to predict outcome. This is true up to Relativity. In the case of Quantum Mechanics, prediction with certainty is quite impossible – the best we can manage is probability. But with Chaos, cause is linked to initial conditions and not based on history. Once the butterfly wing has flapped we can't go back and repeat. We can only concern ourselves with the future, predicting when unpredictability will occur, thereby allowing us to respond appropriately.

As we look back over our series of proponents of scientific mythology, I ask you to consider that, starting with Pythagoras and his numbers and moving up to the present we might be able to ascribe a quality of progression whereby the series of mythic expressions, moving up to the present become less material and more abstract. Given this possibility, for the moment I ask you to humour me as we move into another possibility.

All current science proceeds using the rules of matter, in increasingly abstract forms, with the unshakable assumption that matter exists. Let us just pretend for a bit that the whole universe, what we call "creation," is a complete non-equivocal illusion and what we have always considered as being matter in its many manifestations – air to rock to globular clusters of dark energy has no physical reality except as an apparency created by mind. (See p. 58-9) If that were true what might it mean? Might Creation be nothing more than a great collective metaphorical Maya that we assume is shared by all life forms? This is founded on an understanding and experience that the whole of creation is unified, like all the drops of water in the ocean, there is no separation. Its all one, and on the absolute level, all the same stuff. As a result subject/object relationships can't exist. Everything is already joined. Of course, for most of the universe, ignorant of this possible reality, the apparent reality is diversity, separation, and everything carries on as if that WAS the truth. Its only reality is illusion, Maya – an extension of Plato's story of the shadows on the cave walls.

Does this description of Vedic Mythology satisfy Godel's Incompleteness Theory by being outside every other system? If it does, I have to ask if therefore it could be a key to the fabled "theory of everything"?

To suggest that there is a Mythology of Science, presumes there is a story, common to all branches of science. Even though my luncheon friend will most likely say "bunk! Not science"– I am not dissuaded from my early opinion that, yes there is such a common story. It is a meta-myth that underlies all propositions and claims that, "TRUTH EXISTS and CAN BE DISCOVERED". Whether with the objectively or subjectively based methodology, that fundamental belief keeps all science going. Whether the verification of a "truth" is achieved through experimentation or reasoning, you can rest assured that the two methodologies will not support each other. At its simplest, this is a polarity of "mind or heart". The objective stance gets stuck in a monist position – "there is only our reality". If it cannot be tested and measured its not science, – its only speculative philosophy with no sureness of fact or truth. This is the way "pure"western science has operated since its beginning.

However, I feel that this idol has feet of clay. The question must be asked, "What truth?"–or more appropriately – "Whose truth?"– and only one? – or if more than one, should they be described as "truths". That question must be explored. I don't think that I am alone in having loosely connected the idea of truth with absolute universal verities. Now I am not so sure. Except possibly in some myths, or in

Nature's Law, only known to God, they do not exist for us. Our laws, in most cases, are more pedestrian – and not in that sort of absolute form. If they had been, pursuit of the grand unified field might have been more successful. John Horgan, in "The End of Science" stated that "scientists sacrifice the notion of absolute truth, so they can seek the truth forever."

Not surprisingly, I am left with the "truth" that all truth is relative and recursive. We have generally assumed an ontological value for truth, something truer than "truth" when that isn't what is meant except by ideologies. It's much simpler. In science, truth is claimed when, relative to certain initial conditions being maintained, certain out-put results from certain input. On this basis, the "truth" becomes prediction of output. Given that it is a closed linear system, that would be expected. Certain of the results having been replicated sufficient times, may be elevated to the class of laws- and hence "laws of nature" It would seem that, getting to that level, i.e, declaring a truth, though seen by science as linear, is in fact a chaotic process that will produce its own surprises. The influence of non - linearity has generally been seen as "unwelcome" noise and ignored. However, many more see Chaos as being useful in many parts of science.

And then there is the subjectivity based camp, those like myself whose "truth' is fundamentally based on our personal belief and we might add, intuition. We might even find others that have similar views, based on a similar, but not identical subjectivity. This approach is definitely non-linear. I clearly understand how science must respond.

Mike put it beautifully – "Science has always been an in-house inductive fight over the value of deductive logic"- and it will most likely continue down that path.

What did I learn? – the next time there is a desire to designate a truth, I would remember that I was looking at a metaphor, a translation of a belief system, a closed system with limits. Also I would bear in mind that even if science says its so, doesn't make it my truth -unless that is also my choice – and your is yours!

Reading List

"The works of Plato", Irwin Edman, ed., The Modern Library, N.Y., 1928

"Aristotle for Everybody", Mortimer J. Adler, Bantam Books, N.Y, 1978

"Sophie's World"Jostein Gaarder, Berkley Books, N.Y., 1996

"A Philosophy of Science for Personality Theory"J.F. Rychlak, Houghton Mifflin Co., 1968

"The Universe and Dr Einstein", Lincoln Baarnett, Bantam Books, N.Y. 1950

The Dancing Wu Li Masters", Gary Zukav, Bantam New Age, N.Y. 1979

"The Whispering Pond"Ervin Laszlo, Element, Rockport, 1996.

"The End of Science", John Horgan, Broadway Books, N.Y., 1996

"Chaos", James Gleick, Viking, N.Y. 1987 (source of photocopy illustrations)

"Turbulent Universe", John Briggs and F. David Peat, Harper and Row, N.Y., 1989

"Divine Harmony", John Stromeier&Peter Westbrook, Berkeley hills Books, Berkeley, 1999.

"Chaos"– Theory in the social Sciences, Foundations and Applications", L. Douglas Kiel& Evel Elliot, ed., University of Michigan Press, 4th Edition, 2004.

www.ingramcontent.com/pod-product-compliance
Lightning Source LLC
Chambersburg PA
CBHW051540170526
45165CB00002B/817